D0938971

The Niche in Competition and Evolution

The Niche in Competition and Evolution

by

WALLACE ARTHUR

Department of Biology
Sunderland Polytechnic

A Wiley–Interscience Publication

JOHN WILEY & SONS
Chichester · New York · Brisbane · Toronto · Singapore

Library of Congress Cataloging-in-Publication Data:

Arthur, Wallace.
The niche in competition and evolution.
'A Wiley–Interscience publication.'
Bibliography: p.
Includes indexes.
1. Niche (Ecology) 2. Competition (Biology)
3. Evolution. I. Title
QH546.3.A78 1987 574.5 87-10442

ISBN 0 471 91615 3

British Library Cataloguing in Publication Data:

Arthur, Wallace
The niche in competition and evolution.
1. Population biology
2. Population genetics
I. Title
575.1′5 QH352

ISBN 0 471 91615 3

Typeset by Acorn Bookwork, Salisbury, Wiltshire
Printed in Great Britain by St Edmundsbury Press,
Bury St Edmunds, Suffolk

This book is dedicated to the inspiration of
Georgyi Frantsevich Gause
1910–1986
and his pioneering work in experimental ecology

On the one hand—
'The concept of the niche pervades all ecology'

Pianka (1983)

'The niche concept remains one of the most confusing, and yet important, topics in ecology'

Root (1967)

And on the other—
'The concept of ecological niche will probably turn out to be unnecessary'

Margalef (1968)

'I think it is good practice to avoid the term niche whenever possible'

Williamson (1972)

Contents

PART III COMPETITION, COMMUNITY STRUCTURE AND EVOLUTION

Preface

This book is primarily about interspecific competition. I discuss the population dynamics of competition in Parts I and II; and then move on to a consideration of its evolutionary consequences and its role as an agent of community structure in Part III. Throughout the first two parts, I emphasize the analogy between competing species within a guild and competing genotypes within a population; and I include a brief coverage of the within-species equivalents of the phenomena of competitive exclusion and stable coexistence. I do this largely to indicate to those ecologists without a genetic background that their studies apply to a wider domain than that to which they normally restrict their attention. To give an example: niche differentiation may be a cause of stable coexistence of species *and* of genotypes—because it is a stabilizing mechanism whose operation is not dependent on whether the competing entities interbreed.

I hope that this book will be useful to all ecologists from the 'senior undergraduate' upwards. It is intended to fill the gap between the detail of the primary literature and the necessarily brief and general treatment of competition given in the chapter or two devoted to it in most general ecology texts. An undergraduate reader who has taken only introductory courses in ecology will probably be best reading the book in its entirety. A researcher in population ecology will be able to use it more selectively, and could, for example, go straight to the chapters where I say something new or discuss a topical issue—especially Chapters 5, 7, 8 and 9. As a book that is focused on interspecific competition, and includes only enough basic population genetics to make clear the analogy between competition and polymorphism, I do not envisage it as being aimed at population geneticists—except those interested in coevolution, who should find Chapter 7 relevant, and of course those interested in learning some ecology, who should read the whole book!

The book has several idiosyncrasies, which I should point out to the reader at the outset. First, as the title implies, the concept of the niche is central, and in a way my aim is to give an exposition of basic niche theory in population and evolutionary ecology. Mechanisms of, for example, competitive coexistence, other than those that can be formulated in niche terms, are mentioned but not discussed in detail. This of course leads to the question of what exactly is a niche—which is dealt with in the first chapter. Second, I devote considerable space to the laboratory-based, experimental approach. This is currently out of favour in population ecology, as lamented by Mertz and McCauley (1982). However, it seems to me essential that we persist in our attempt at a thorough understanding of simple systems while we are trying to gain a foothold in the study of complex natural communities; and our knowledge of 'simple' systems—such as Gause's *Paramecium*—is not nearly as complete as is sometimes thought. Third, the examples given are predominantly (though not exclusively) zoological. While this bias may seem undesirable, I can at least take comfort from the fact that there is a fairly recent book on competition in plant communities (Tilman, 1982).

The role of competition both as an agent of community structure and as an agent of evolutionary change has recently come in for considerable scrutiny; and the debate between those who attribute an important role to competition and those who do not has become quite intense. I deal with this recent controversy in Part III of the book; Parts I and II deal with the population dynamics of competition in systems where its importance is usually not at issue.

While much of this book is taken up, as might be expected, with reviewing previous work, three of the chapters contain material that is largely new. In Chapter 5, I describe the 'mechanistic competitive exclusion principle', which is intended to be both an advance over earlier versions of the principle and a reasonable generalization about competing species in nature. In Chapter 7, I offer an approach to case studies of competitive coevolution which should prove less prone to premature claims of discovery of this phenomenon than were earlier approaches. In Chapter 9, I make the proposal that there is a general correspondence between important evolutionary changes and a *lack* of competition.

Several people read and criticized part or all of the manuscript, and I am grateful to them for their helpful comments. Bryan Clarke, John Lawton and Amyan Macfadyen commented on an earlier manuscript, material from which has ended up, in revised form, in Chapters 2 and 5. Michael Hassell and Paul Mitchell read through the complete typescript and made numerous helpful suggestions. I am particularly grateful to John Lawton for suggesting that I tear up my earlier explanation of the mechanistic competitive exclusion principle, which in retrospect I agree was undecipherable—it is for the reader to decide whether what I replaced it with is as good as it should be. Finally, I thank Susan Harrison (typing) and Michael Dixon and colleagues (at Wiley) for their usual efficiency at their respective stages in the transmutation of an illegible scribble into print.

Chapter 1

Introduction

1.1 GENERAL INTRODUCTION

The central theme of this book is interspecific competition. I examine first the population dynamics of such competition, together with the analogous population dynamics of competition among genotypes *within* a species (Parts I and II). Following this, in Part III, I discuss the questions of how common interspecific competition is in nature, what sort of coevolutionary changes (if any) it gives rise to, and what effects both competition and coevolution may have on the structure of natural communities.

Parts I and II are particularly concerned with attempts to view the outcome of competition among species or genotypes in terms of the niche. I deal first with situations in which there is no stable equilibrium, namely competitive exclusion and its intraspecific equivalent, the fixation of one allele at the expense of another. I then move on to stable systems, concentrating on those where the stability appears to arise from some form of niche differentiation—either among species, giving rise to stable coexistence, or among genotypes, producing balanced polymorphism of the 'multiple niche' variety.

Current attitudes to multiple-niche coexistence and polymorphism within (respectively) ecological and population genetics theory are markedly heterogeneous. In the former case, we have moved from a situation where coexistence caused by niche differences was almost universally considered to be common in nature to a situation where many ecologists consider it to be rare, and hence unimportant, while others continue to hold the 'conventional' view. Some ecologists have questioned the commonness of niche-based coexistence on the basis that competing species may often coexist by alternative means (e.g. Koch, 1974a,b). Others have made an even more fundamental assault on niche theory by proclaiming that natural populations are very rarely resource-limited (Pimm, 1980, 1982, 1984). If true, such an assertion would relegate *all* mechanisms of competitive coexistence to a relatively minor role in ecological communities, since, under this view, competition itself is rare (depending on how you define it: see next section).

The controversy between the resource-limitation and predator-limitation schools will be considered in Part III. Parts I and II embark upon an examination of the population dynamics of interspecific competition *in situations where it is known to be taking place*—principally in laboratory populations. The separation of these two issues is essential. However common or rare competition is in nature—which must be determined through field studies—it clearly does occur sometimes. If we want to understand the structure of those communities in which competition does occur, we need to know something of competitive population dynamics. Such a knowledge is built up from, among other things, perturbation of the relative numbers of competing species to see if their coexistence is stable. While such experiments can in principle be conducted in the field, the practical problems encountered are usually prohibitive; and thus the detailed analysis of competitive population dynamics is best approached through a study of laboratory populations.

In addition to stressing the need to separate the two questions of how frequently competition occurs, and how it 'works' when it does occur, I think it is necessary to argue for the inclusion of the latter, and the laboratory-based approach that is involved. There is currently a rather 'anti-laboratory' mood in ecology, as has been pointed out by Mertz and McCauley (1982), which is rather unfortunate. The idea that ecology can progress solely through a mixture of theoretical models and field studies, and without the detailed investigation of population dynamics that multi-generation experiments in the laboratory make possible, is a very blinkered, and in my opinion erroneous, view. Nor can we pretend that the experiments of earlier workers such as Gause (1934, 1935) and Park (1948, 1954) have told us all we need to know about the population dynamics of interspecific competition. The advances made by these, and other, experimentalists were enormous, but, as we shall see, many questions remain unanswered.

The most important of these questions is whether it is possible to state a general set of conditions which must hold if coexistence is to ensue; and if so, what those conditions are. This, of course, is the realm of the principles of competitive exclusion and limiting similarity. These 'principles', which are not yet sufficiently well tested to deserve that name, can serve either as useful hypotheses or as worthless terminological muddles, depending on exactly how they are stated. I will leave this issue until the end of Chapter 5, as it does not make sense to deal with generalizations about competition before we have looked at some individual case-studies. For the moment I will only comment that Pontin's (1982) writing off of the competitive exclusion principle as 'not worth saying' and limiting similarity as an 'unrealistic idea' is somewhat short-sighted. We should not so readily discard what might become our most important generalizations simply because they are sometimes badly worded.

Turning to the status of multiple-niche polymorphism in population genetics theory, we find an altogether different state of affairs from the above. Here, the main controversy is not about whether polymorphism is common in natural populations—such commonness has not been disputed since the classic studies of Lewontin and Hubby (1966; see also Hubby and Lewontin, 1966 and Harris, 1966) on enzyme polymorphism. Nor is there a heated debate between supporters of the concept of multiple-niche polymorphism on the one hand and some alternative camp. Rather, the main debate is between supporters of the view that most polymorphisms are stabilized by some form of balancing selection, of which that resulting from niche differences is only one, and the alternative view that the majority of polymorphism, at least at the molecular level, is 'neutral' and not subject to selection at all (Kimura, 1968, 1983). Thus while multiple-niche polymorphism is controversial both in the sense that its commonness in nature remains to be established, and in the sense that it is involved, albeit cryptically, in the still-unresolved debate between selectionists and neutralists, it does not occupy such a central position in population genetics theory as multiple-niche coexistence does in the theory of interspecific competition.

The 'genetic' component of Parts I and II parallels the ecological one, in that we move from a consideration of what happens in competition between genotypes with essentially the same niche (Chapter 3) to the outcome of competition among those whose niches are different (Chapter 6). The basic ideas about production of balance through niche differences, in both competitive and polymorphic situations, are treated together in Chapter 4. I am particularly concerned to emphasize the parallels between interspecific and intergenotypic competition, an emphasis which in the past has been largely lacking.

The main difference between the two situations is that, in competition between genotypes, the different forms interbreed. The most significant result of this interbreeding is that it introduces an additional balancing mechanism—namely heterozygous advantage. There is no parallel mechanism of 'hybrid advantage' in competition between species in cases where they interbreed, because, of course, interspecific hybrids suffer from a varying (and usually severe) degree of *dis*advantage. How big a difference this constitutes between interspecific and intergenotypic competition depends on how common heterozygous advantage turns out to be in the latter process. The fact that sickle-cell anaemia (Allison, 1955) and a handful of other cases repeatedly turn up as examples of heterozygous advantage suggests that this phenomenon is fairly rare; though the opposite has often been argued.

The relevance of interspecific competition for evolutionary theory is not confined merely to parallels between its population dynamics and the dynamics of polymorphism. Competition can, like predation, act as a selective agent on the interacting populations, one or both of which (or all, in multi-species situations) may thus evolve in response to their competition. Where the selective agent on one population is another population (of a competitor or predator), the selective agent itself may evolve, and the situation is thus one of *coevolution*. This contrasts with selection caused by abiotic agents such as temperature, where the relationship between selector and selected is unidirectional.

As with competition, there are two quite separate questions about competitive coevolution, namely (a) how frequently does it occur in nature? and (b) what form does it take when it does occur? Although ideally the first of these questions would be approached through field studies and the second by laboratory work, several factors, principally the long time-scale involved, make production and characterization of coevolutionary changes in the laboratory very difficult. Consequently, much attention has had to be paid to observation of phenotypic changes occurring in nature at borders between allopatry and sympatry. The interpretation of such changes is fraught with problems, as has frequently been pointed out (Grant, 1972; Arthur, 1982a); and proposed coevolutionary patterns based largely upon them, notably character displacement, rest on a very shaky observational foundation.

Even if character displacement (or some other coevolutionary process) does sometimes occur when two closely related species compete, it need not be a widespread process with much effect on the overall structure of natural communities. The question of whether or not 'community-wide character displacement' exists has become very controversial, together with the 'null-model' approach now sometimes used to address the question (see Strong *et al*., 1979). This issue will be discussed in Chapter 8, after the evidence for particular cases of character displacement has been examined in Chapter 7.

While there is much debate about the extent to which competition is an important factor causing evolutionary change, something that receives little attention from evolutionary ecologists is the question of whether lack of competition may have some significance in evolution. The importance of such a lack is most often proclaimed by students of long-term (macro- and/or mega-) evolution, e.g. Simpson (1944), Stanley (1979), Raff and Kaufmann (1983) and Arthur (1984), usually in connection with the origin of a new 'type' at anything from the species to the phylum level. Concentrating, as they do, on microevolutionary changes, evolutionary ecologists tend to see lack of competition as unimportant or irrelevant. In Chapter 9 I attempt to persuade them otherwise.

The above discussion has served to introduce the topics and controversies with which this book will deal, and to show the broad sequence in which they will be considered. The remainder of this introductory chapter will be devoted to clarifying three areas of potential confusion. After this has been done, we can, I hope, progress to examine the real scientific issues at stake without these being obscured by a sort of terminological haze.

1.2 ON DEFINING INTERSPECIFIC COMPETITION

Various ecologists have proposed verbal definitions either of competition in general or of interspecific competition in particular (Birch, 1957; Bakker, 1961; Milne, 1961). These present difficulties, however, both because no two definitions are exactly the same, and because any one of them contains words whose meanings are themselves obscure. I sympathize with Williamson (1972) who objected to one verbal definition (that of Milne, 1961) because it replaced a single undefined term (competition) with seven. The alternative approach is to define competition symbolically, in terms of the direction of effect each species has on the other's population—inhibitory (−), stimulatory (+) or neither (0). If we restrict our attention for the moment to two-species situations, competition is definable as a (−, −) interaction, while other types of interaction are represented by other combinations of symbols—e.g. predation is (+, −). This sort of system goes back at least as far as Haskell (1947), who saw it in very general terms. It was co-opted into ecology by Odum (1971), and made a little more precise by Williamson (1972) who noted that the pluses and minuses referred specifically to the direction of effect on equilibrium population size or on rate of population growth.

There are several advantages in adopting the (−, −) definition of competition, one of which is that it connects much more readily with relevant models, such as the well-known Lotka–Volterra model (see Chapter 2), than do verbal definitions. However, it is becoming increasingly apparent that different approaches are required for situations in which the (−, −) effect is achieved in different biological ways. There are at least three quite distinct ways of producing the mutually-inhibitory effect: through use of a common limiting resource, through some form of direct attack—physical or chemical—on the alternative species and through the activities of a predator consuming individuals of both species. These three situations may be referred to as 'exploitative', 'interference' and 'apparent' competition respectively, the last category having only been named relatively recently (Holt, 1977, 1984). These distinctions are highly relevant here, because I will be concentrating, throughout, on exploitative competition—though it is necessary to make the proviso that it is sometimes not clear in a particular competition experiment (e.g. Gause, 1935) whether the interaction involves exploitative or interference effects or both. Having declared this at the outset, I will feel free to use the term 'competition' to mean exploitative competition except where I specifically state otherwise.

Whether this definition can be easily extended to a multi-species context depends on whether 'higher-order interactions' occur. An example of such an interaction is where the effect of species A on species B is altered by the presence of species C. If there are no higher-order interactions, then a multi-species situation in which all pairwise interactions are (−, −) represents competition. If there *are* higher-order interactions, and specifically those of a kind where the *direction* rather than just the magnitude of the effect of one species on another varies with the introduction of a third species, then it is difficult to know what to call the overall interaction. At the present elementary state of our knowledge of population interactions, such complex situations are best avoided altogether.

Finally, I should point out (as Birch, 1957, has done) that population geneticists, when discussing 'competition' among genotypes, normally use the word in a broader sense than the intergenotypic equivalent of exploitative competition; indeed they sometimes equate competition between genotypes with natural selection. I will return to the problem of defining this kind of competition in Section 1.4.

1.3 THE NATURE OF THE NICHE

Since we will subsequently be looking in detail at the idea that stable coexistence and stable polymorphism can result from differences in the *niches* of the competing species/genotypes, and since the term 'niche' has been used in several distinct ways, it is essential that we start by restricting the meaning of 'niche' to something fairly specific. The criterion for deciding which specific meaning should be attached to the term cannot be correctness, since there are too many alternative usages, and since the term has no clear origin in ecology. (It seems to have filtered in gradually from general usage, via the work of Johnson, 1910, and Grinnell, 1914, 1917—see Hutchinson's (1978, Chapter 5) history of the niche concept.) Rather, the criterion must be usefulness to the task at hand. Thus, the usage of 'niche' developed below, which is to be related to competition, would be expected to differ from the usage adopted in a book on, say, geographical distribution. Of course, it would be better to have entirely different words for the different niche concepts, but history has not been kind to us in this respect.

The three authors who gave fairly concrete formulations of the niche were Elton (1927), Hutchinson (1957, 1965) and MacArthur (1968, 1970, 1972; see also May and MacArthur, 1972). Elton defined the niche of an animal as 'its place in the biotic environment, *its relations to food and enemies*'. Hutchinson (1965) described the niche as a hypervolume, each of whose dimensions 'corresponds to a relevant variable in the life of a species' and whose boundaries were defined by the species' tolerance limits on those dimensions. MacArthur's niche is the resource utilization function (RUF) which is a plot of utilization against some quantitative resource variable (e.g. size of seed eaten by granivorous birds; see Figure 1.1). Both Hutchinson's and MacArthur's niches are formal concepts, being based in set theory and normal distribution theory respectively, while Elton's niche is without such a formal basis. On the other hand, Elton's niche shares with MacArthur's a concentration on biotic variables, while Hutchinson's includes both biotic and abiotic factors. If we make only these two distinctions (formal/non-formal and biotic/biotic+abiotic), then there are four possible niche concepts, represented by A to D in Figure 1.2. Under this scheme, A represents

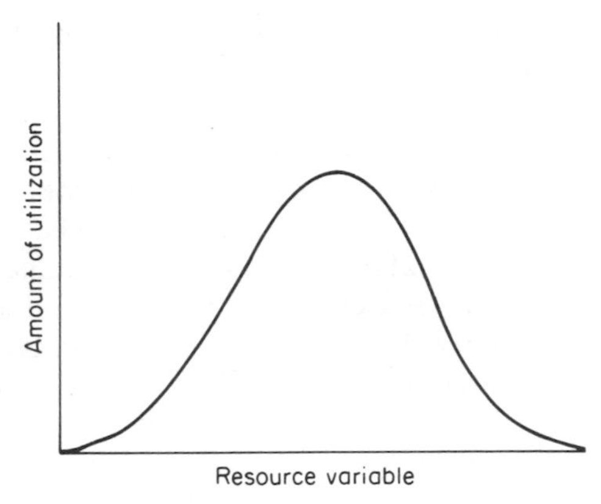

Figure 1.1 A resource utilization function (RUF)

MacArthur, B Elton, and C Hutchinson. Although Grinnell never defined his view of the niche, it was clearly non-formal. Also, since he discussed it in a paper where biotic and abiotic factors were distinguished without specifically attaching it to either (Grinnell, 1924), I suspect that he viewed it as all-inclusive in this respect. If my interpretation of his views is correct, D in Figure 1.2 represents Grinnell, though I must agree with Hutchinson (1978) when he says that Grinnell used the term niche 'without a completely clear indication of its meaning'.

Since I want to adopt a fairly quantitative approach, it will be necessary to use one of the formal versions of the niche concept. The choice, then, is between a Hutchinsonian hypervolume and MacArthur's RUF. I will use the latter, for the following reasons. First, although RUFs can be made multi-dimensional (see Pianka, 1981, 1983 Chapter 7), and hypervolumes can be simplified to some extent by omitting some of the less important dimensions, the basic idea of an RUF is expressible with just a single resource axis, while the whole essence of a hypervolume is that it is constructed in many dimensions. An approach which attempts to see things in a single dimension if possible but allows us to make the picture more complex when necessary seems infinitely preferable to an approach which at the outset launches us into multidimensionality. Also, the concentration on a *resource* axis that is

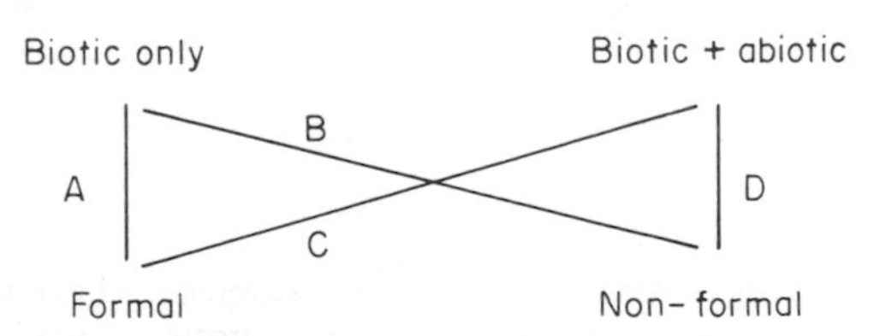

Figure 1.2 Four possible niche concepts

embodied in the RUF concept is precisely what we need when dealing, as we are here, with exploitative competition.

Another advantage of the 'RUF-niche' over the 'hypervolume-niche' is that while the latter is only concerned, along any particular dimension, with the tolerance limits of the organism under consideration, the RUF concentrates on what happens *between* the tolerance limits. That is, it concentrates on the pattern of utilization along a resource axis rather than just with identifying the section of the resource axis that gets used at all. It is thus a more intrinsically *behavioural* concept, reflecting what organisms actually do, in terms of resource use, rather than just dividing a resource axis into a section that is utilizable and flanking sections that are not.

Having decided to use 'niche' to mean RUF, we now need to consider the question of what type of entity may be described as having a niche. Perusal of the literature reveals cases of organisms, genotypes, phenotypes, populations, species and environments all possessing 'niches'. It hardly seems sensible to use the term in such a variety of ways, and it will be helpful if, at the outset, we adopt a more restricted, and more definite, usage. I will consider here the use of the niche in the context of interspecific competition. Niches in the context of polymorphism will be discussed in Section 1.4.

The first distinction that must be made is between organisms (or groups of them) and the environment. Environments are sometimes described as having 'empty niches'. This does not, of course, mean that they have empty RUFs—such a concept is meaningless. Rather, it means that some resource within the environment is unutilized, and that there is thus an ecological role that could be adopted by an immigrating population. In terms of RUF theory, this is equivalent to a section of the resource axis being unoccupied by the RUFs of the existing species—see Figure 1.3. It is clearly useful to have a label for this situation, and 'empty niche' is a problematic one, as we have seen. Some authors have advocated the use of 'niche space' for the environment and 'niche' for organisms. This seems a sensible form of terminology, and it will be adopted here. It should be noted, however, that some authors, e.g. Cohen

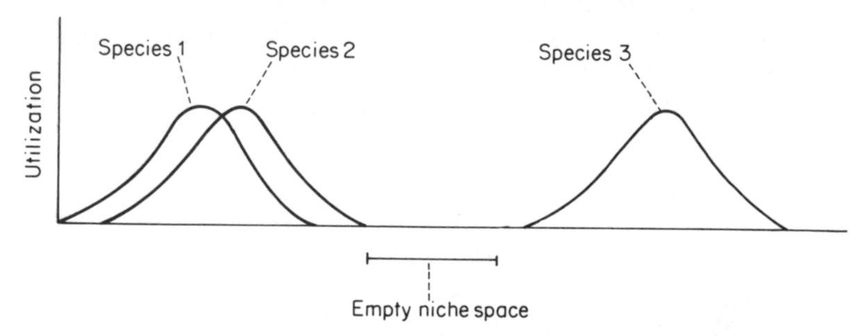

Figure 1.3 Empty niche space in an environment where all points on the resource axis are available for consumption, but not all are consumed

(1978) have used 'niche space' in a different way and that some (e.g. Crozier 1974) have employed terms like resource landscape in essentially the same way as I here use 'niche space'. Thus the situation depicted in Figure 1.3 can be described as empty niche space; it will be available to immigrants whose niches correspond to it. Of course, 'space' here refers to resource dimensions, not physical ones.

Having restricted 'niche' for use with organisms rather than environments, to what level(s) of grouping is it applicable—individual, population or species? Let us imagine a hypothetical situation here, namely an artificial environment where the food supply consists solely of a single type of seed, and the seeds exhibit continuous variation in size, from very small to very large. If we introduce a single individual of a granivorous bird species into such an environment, we may observe its utilization of different seed sizes, and so, after a time, describe its RUF. If we introduce a population rather than an individual, we may do likewise. In this case, we would expect the RUF to be broader (i.e. to have a greater variance) because of variation in morphology and behaviour among the birds themselves. Finally, were it possible to gather up all extant individuals of the species concerned and introduce them into our hypothetical environment, we could again produce an RUF, and this would be broader still, for the species as a whole will exhibit more variation than any one of its populations.

The above scenario shows that RUFs can be used at any of the three levels mentioned, but this does not necessarily mean that they should be so used. A problem arises, because the observed RUFs will depend on the nature of the environment. For example, if the seed size range in our hypothetical environment is restricted so that there are no seeds larger than the optimum size for our species of bird, the RUF (whether of individual, population or species) will be altered in range and variance and will be highly skewed to the left. If we wish to actually measure RUFs in reality, therefore, we need to do so in a particular environment, and the pattern we obtain there may not apply in other situations. Since it is populations, rather than species or individuals, that inhabit particular natural (or artificial) environments, it seems most sensible to consider the RUF to be a characteristic of a population; and indeed of a particular population–environment combination. This means that the RUF has similar properties to heritability in quantitative genetics: it may differ for different populations in the same environment or for the same population in different environments.

The specific use of the niche to refer to populations has the advantage that we can begin to analyse it, i.e. to determine what sources contribute to the variation in diet that it represents. A good start in this direction has been made by Roughgarden (1972, 1976) who has distinguished the within- and between-phenotype components (WPC and BPC) of the niche. These are components due, respectively, to the behavioural flexibility of individuals and to differences between individuals. This seems a useful distinction, and I will return to it later. It should be stressed, though, that differences between sexes

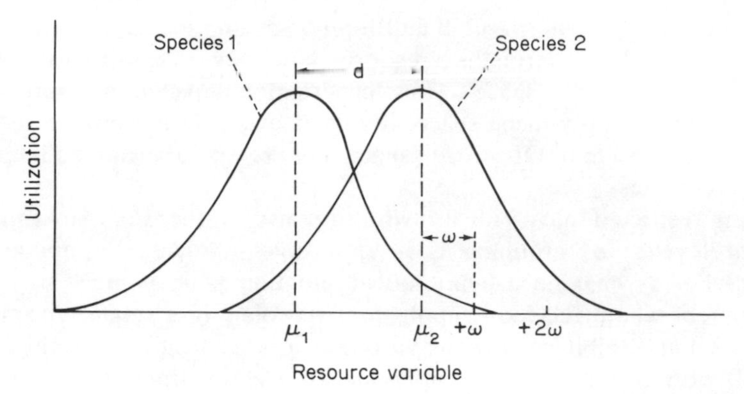

Figure 1.4 Definitions of some niche statistics. μ—Niche location. w—Niche breadth (standard deviation). d—Niche separation (absolute). d/w—Niche separation in relation to breadth. (This is an inverse measure of niche overlap)

and between age-classes will usually far exceed those between phenotypes within a particular age–sex category, except, perhaps, in species with a 'major' polymorphism (see next section).

Adopting a view of the niche as the RUF of a particular population in a particular environment allows us to attach precise meanings to the sometimes vague concepts of niche location, breadth, separation and overlap. These meanings, which derive from May and MacArthur (1972) and other earlier authors, are illustrated in Figure 1.4. In addition to being precisely defined, niche parameters such as d and w can be measured and hence niche separation or overlap (d/w) determined in appropriate experimental situations (e.g. Arthur and Middlecote, 1984a). Clearly, experiments of this sort are vital if theoretical predictions of the limits to niche overlap compatible with coexistence are to be tested.

1.4 POLYMORPHISM AND THE NICHE

The concept of the niche became relevant to studies of genetic polymorphism when Levene (1953) showed that it was possible to have stable polymorphism as a result of niche differences between competing genotypes. No other balancing mechanism—such as heterozygous advantage—was necessary for Levene's model to work; niche differences were sufficient in themselves to produce the state of balance, just as they are in cases of species coexistence.

The basic situation we have to picture here (which will be developed in more detail in Section 6.1) is that of a polymorphism involving alleles A_1 and A_2, and consequently the genotypes A_1A_1, A_1A_2, and A_2A_2. If there is no dominance of one allele over another then we have three separate RUFs. However, if there is complete dominance (say of A_1 over A_2), then the number of RUFs reduces to two. This is because, in the context of polymorphism, an RUF is a property of a *phenotype*, not a genotype. Clearly, in more complex polymorphisms involving more than two alleles, there will be correspondingly more RUFs to be considered.

The restrictions placed on the use of niche/RUF in the previous section still apply when we are considering multiple-niche polymorphism. That is, the niche refers not to individual organisms, species or habitats, but to a particular population in a particular environment. All we are now doing is adding the complication that there may be two or more subsets of the population whose RUFs differ significantly, and which should not, therefore, be lumped together.

Although such subsets may occur for non-genetic reasons—particularly in the case of populations with different age-classes or life-stages—we will concentrate here on genetically-based differences between niches. Probably the most widespread of these is sexual niche dimorphism. To what extent other known polymorphisms are accompanied by niche differences is not clear, as relatively few studies have been carried out in this area (but see Chapter 6). However, it seems likely that some kinds of polymorphism will more often be accompanied by niche differences than others. A working classification of types of polymorphism is given in Figure 1.5, along with some well-known examples belonging to the various categories.

It is important to stress that category (3) in Figure 1.5 is not defined so as to exclude the possibility of niche differences. The controversy about enzyme polymorphism is centred on whether functional differences between genotypes are (a) negligible or (b) slight but significant. Niche differences may thus turn up in the context of enzyme polymorphism, but they are likely to be difficult to detect in comparison with those among the different phenotypes belonging to a 'major' polymorphism such as the colour or banding polymorphism in *Cepaea nemoralis*. Even in the case of *Cepaea*, we do not yet know whether the different phenotypes have different RUFs, though behavioural differences between them have been identified (Jones, 1982).

Turning to the question of what constitutes competition among genotypes (or, more correctly, phenotypes), Birch (1957) has pointed out that such competition need not be for a limited resource. This is certainly true. A population may be limited by the density-dependent action of a predator or parasite, and yet a new phenotype may arise (by mutation or immigration) and sweep through a population, ultimately replacing the 'old' form completely. In a very real sense, the old phenotype has been outcompeted, even though resources of food or space never became limiting. However, it is important to recognize that the population did have a limiting *factor* (a broad category subsuming that of 'limiting resource'), and that the competing phenotypes shared that limiting factor. It is such sharing that constitutes competition and that ultimately ensures the extinction of the less fit phenotype (if no balancing forces are acting). As Ford (1971) pointed out, during a phase of overall population expansion, the subpopulation of an unfit

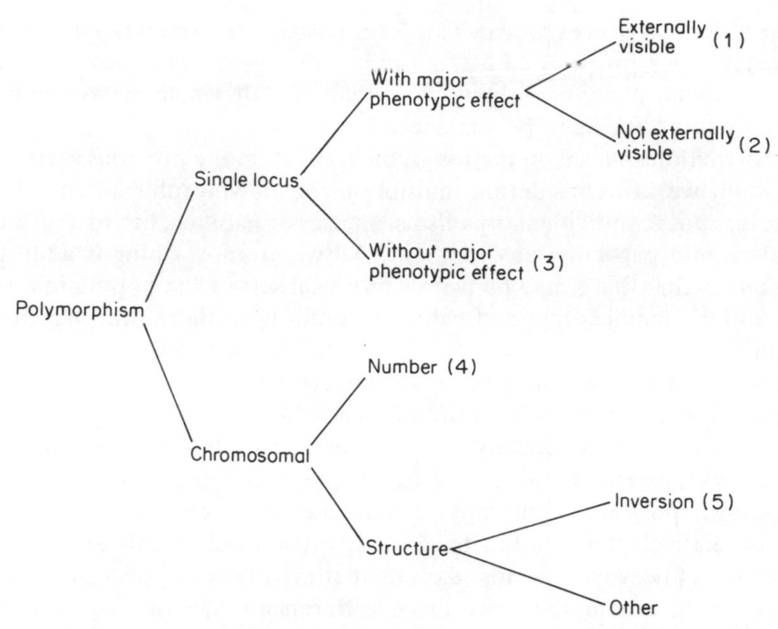

Figure 1.5 A working classification of types of polymorphism.
Examples: 1. Pigmentation polymorphism in *Biston*, *Cepaea*.
2. Sickle cell anaemia in man.
3. Allozyme polymorphism in general.
4. B-chromosome polymorphism in grasshoppers.
5. Inversion polymorphism in *Drosophila*

phenotype may itself increase in size despite its comparative unfitness because, temporarily, the population is not being limited and the genotypes are not in competition.

1.5 THE DISAPPEARING NICHE

So far in this chapter I have said quite a lot about the niche. Indeed, the niche is one of the central themes of the book as a whole. However, while the concept will re-surface in Part II, it will, before then, largely disappear. The reason for this is that students of directional change in populations have mostly failed to quantify the degree of similarity or difference in the niches of the competing forms. The reason for this omission is probably different in the two types of system with which I deal here—namely polymorphism within species and competition between them.

Those workers studying competitive exclusion have tended to assume niche similarity rather than demonstrating it. This is problematic, since if you do not look for niche differences you will not find them, and if they are only looked for subsequently to the finding of a state of stable coexistence, an erroneous link between coexistence and niche differences could easily arise. I will return to this problem later. For the moment I will merely make the point that since students of competitive exclusion have not provided quantitative data on niches I will have relatively little to say about the niche in the following chapter. We thus proceed with the uneasy assumption that in the experiments concerned the niches are, in some sense, the same.

The lack of quantification of niche parameters in studies of directional selection has a different explanation. In fact, niche statistics are absent in *most* studies of polymorphism, including those that appear to be stable. The omission of niche data in many cases of stable polymorphism would appear to be due to a general feeling among some groups of population geneticists that niche differentiation is relatively rare as a stabilizing mechanism compared, for example, to heterozygous advantage. Given such a feeling, there is even less motivation for the population geneticist to look for niche differences in situations where one allele is replacing another than there is for the ecologist to seek them in a system of competitive exclusion, because not only is the expectation a negative result, but the finding, whether positive or negative, has no direct bearing on what is taken, rightly or wrongly, to be the 'main' agent of stability. This is another point to which I will return later. For the moment, as in the interspecific context, we proceed with the assumption that competitive elimination of one allele by another implies that, in some sense, the niches of the competing entities are the same.

PART I

DIRECTIONAL CHANGE IN COMPETITION AND POLYMORPHISM

The basic message of this part of the book is that if two different types of organism compete, then in the absence of any stabilizing mechanisms—of which niche differentiation is one—the inevitable result is the extinction of whichever type is less 'fit' or 'competitively able' under the prevailing set of environmental conditions. This conclusion holds regardless of whether the organisms interbreed—that is, the 'types' are phenotypes and the phenomenon is directional selection leading to fixation—or not, that is, the 'types' are species and the phenomenon is interspecific competition leading to competitive exclusion. The conclusion is also independent of the mechanism of limitation of the populations concerned (whether by resources or predators) so long as the competing types are indeed limited by the same factor. This last comment is almost tautological, since the sharing of a limiting factor is one of the many definitions of competition.

The only qualification that needs to be made to this very general, and almost axiomatic, process of competitive replacement is that the 'different types of organism' that are competing do need to be significantly different. Where the difference is negligible, there is no deterministic force causing competitive elimination of either form, and the fate of the system is then in the hands of stochastic processes. While this is unlikely to be the case in competition between species, or in the case of 'major' polymorphisms, it may be a widespread state of affairs in the case of allozyme polymorphism, where the differences between the genotypes are slight.

Chapter 2

Competitive Exclusion

2.1 BASIC THEORY

At about the same time as Grinnell (1924) and Elton (1927) provided us with early formulations of the niche, Lotka (1925) and Volterra (1926) developed the first general theory of interspecific competition. While many additional models of competition have since been developed, some as extensions of the Lotka–Volterra model (e.g. Strobeck, 1973; Shorrocks *et al*., 1979) and others as quite distinct approaches (e.g. Stewart and Levin 1973), the model developed originally by Lotka and Volterra (together with its discrete-time equivalent) remains the simplest *general* model of competition between two species, and thus continues to occupy a special place in ecological theory. I say 'general', because it is important to note the dichotomy between such general models, which attempt to give us an overall picture of a population process, and the detailed, situation-specific models whose aim is to mimic a particular system with a view, perhaps, to predicting future patterns of growth. Maynard Smith (1974) refers to the general category simply as 'models'; the detailed category he calls 'simulations'. The latter is a rather confusing label, since, as Maynard Smith himself points out, it is not synonymous with simulation as performed on a computer; yet such a type of model would often provide the basis for a computer simulation. For this reason Nisbet and Gurney's (1982) term 'tactical models' seems preferable. These authors refer to the general category as strategic models.

As I am concerned, in this book, with developing a general picture of competition and niche theory, tactical models will not be discussed. The focus, in both ecological and genetic contexts, will be on strategic models, of which the Lotka–Volterra is one. Whether a particular model in this category is useful or not is much less easy to determine, and to achieve consensus upon, than whether a particular tactical model is useful. If a tactical model fails, for example, to predict an 'outbreak' of a pest population and predicts such outbreaks when they do not occur in practice, the model is clearly of little use. However, since strategic models are not intended to be accurate descriptions of any particular system, the fact that some systems fail to obey them does not require their dismissal. In the case of the Lotka–Volterra model, some simple systems happen to correspond to it fairly well, e.g. Gause's (1934), *Paramecium* experiments, while others, such as Ayala's (1969) *Drosophila* experiments, do not. But neither of these facts should lead us to reach any general conclusion on the utility of the model.

Perhaps the most important test of a strategic model is whether it helps us to think about the process being modelled in a quantitative way. The Lotka–Volterra model certainly does this for competition, and it is consequently useful both in its own right and as a quantitative 'connection' with the niche concept which, in the hands of some authors, becomes all too qualitative and vague. The purpose of the present section is to describe the Lotka–Volterra model and to discuss its relationship to niche theory. The description of the model itself will be fairly brief, since such descriptions can be found in many ecological texts; and will be centred on the version of the model given by Gause (1934) rather than the versions given by Lotka and Volterra themselves, which are now less commonly employed.

The Lotka–Volterra model for two species (L–V model from here on) is a simple extension of the logistic, as follows:

$$\frac{dN_1}{dt} = r_1 N_1 \frac{(K_1 - N_1 - \alpha N_2)}{K_1} \tag{2.1}$$

$$\frac{dN_2}{dt} = r_2 N_2 \frac{(K_2 - N_2 - \beta N_1)}{K_2} \tag{2.2}$$

where N = population size, K = carrying capacity, α,β = competition coefficients, and r = intrinsic rate of natural increase. The only difference between this model and the logistic model for a single species is the addition of one component ($-\alpha N_2$ or $-\beta N_1$) to each equation. These components indicate the interaction between the species, and α and β are the competition coefficients whose magnitude reflects the intensity or 'strength' of competition taking place.

It is necessary to pay some detailed attention to the nature of the competition coefficients, and in particular the following points should be noted:

1. Each coefficient represents the inhibitory effect *per individual* of one species on the rate of population growth (and on the equilibrium population size) of the other.
2. They are constants. In practice, this may be true neither among individuals nor between time-periods.
3. If the coefficients both take a value of zero, there is no interspecific interaction and the model simplifies to two independent logistics.

4. In competition, both coefficients are positive. If one coefficient is zero, the other positive, the situation is describable as amensalism.
5. In a situation where stable coexistence is occurring, the coefficients can be calculated as $\alpha = (K_1 - \hat{N}_1)/\hat{N}_2$ and $\beta = (K_2 - \hat{N}_2)/\hat{N}_1$, where $\hat{N}$ indicates equilibrium population size in mixed culture (see Ayala, 1969).
6. In multi-species situations, α and β are replaced by a series of coefficients α_{ij}, where each measures the inhibitory effect of an individual of the jth species on the rate of population growth in the ith species (see e.g. Strobeck, 1973).

Let us now examine the conditions for stable coexistence under the L–V model. What we are actually looking at, of course, is the threshold between conditions resulting in competitive exclusion and those resulting in stable coexistence. This threshold can be expressed either as the condition for coexistence or, conversely, as the condition for competitive exclusion. It is more conventional to use the former label, probably because population biologists have, as Maynard Smith (1978) has put it, 'an obsession . . . with equilibrium situations'; but for the purposes of this chapter it may well be more sensible to think of things the other way round.

It is worth noting that stable coexistence and competitive exclusion (the latter preceded, of course, by a phase of transient coexistence) are not the only possible outcomes of competition under the L–V model. Theoretically, at least, it is possible to have two non-interbreeding entities which have the same values of r, K and α/β. Under these conditions, there is a state of neutral coexistence similar, in its stability characteristics, to a state of neutral polymorphism. However, while the latter phenomenon is the basis for a whole theory of molecular evolution (Kimura, 1983), 'neutral coexistence' has never been seriously considered by ecologists, and there exists nothing in competition theory analogous to Kimura's genetical theories. This 'gap' in the analogy between the theories of coexistence and polymorphism is almost certainly justified—the nature of interspecific differences makes neutral coexistence exceedingly unlikely to occur in nature, while the differences between minor genetic variants such as allozymes are slight enough to urge serious consideration to Kimura's views.

According to the L–V model, stable coexistence of two competing species will occur if and only if:

$$\alpha < K_1/K_2 \quad \text{and} \quad \beta < K_2/K_1 \tag{2.3}$$

If α and β are greater than these values, then competitive exclusion takes place.

The 'rationale' behind these inequalities can be put in the following way (see MacArthur, 1972). Globally stable coexistence can only occur if each species is able to successfully invade an equilibrium population of the other. We will consider the situation of species 1 invading an equilibrium monoculture of species 2. In this situation, if the colonizing propagule of N_1 is very small, we can say that:

$$N_1 \simeq 0 \quad \text{and} \quad N_2 = K_2 \tag{2.4}$$

It will help, for current purposes, to consider the *per capita* growth rates. The L–V equation for species 1 is given below in this form (and with the position of K shifted for convenience).

$$\frac{dN_1/dt}{N_1} = \frac{r_1}{K_1}(K_1 - N_1 - \alpha N_2) \tag{2.5}$$

Now since $N_1 \simeq 0$ and $N_2 = K_2$, we can see that the *per capita* rate of increase will be positive if and only if $(K_1 - \alpha K_2) > 0$; or, alternatively, if $\alpha < K_1/K_2$. By a parallel procedure, it can easily be shown that species 2 can invade an equilibrium population of species 1 if and only if $\beta < K_2/K_1$. Clearly, both of these invasions are required for stable coexistence.

While the inequalities that constitute the conditions for coexistence under the L–V model are very simple mathematical entities, their biological interpretation is rather problematic. This is because the mechanism of competition is not specified in the model. Different biological mechanisms of competition require different interpretations of the conditions for coexistence.

Let us consider the simplest situation, that where the two species have equal carrying capacities ($K_1 = K_2$). The conditions for coexistence then become:

$$\alpha < 1 \quad \text{and} \quad \beta < 1 \tag{2.6}$$

What this means is that, for either species, one of its individual members must have a smaller inhibitory effect on the rate of population growth in the alternative species than on its own species. (Recall that the growth rate of species 1 (for example) is reduced by a factor $(- N_1 - \alpha N_2)/K_1$ so that if $\alpha = 1$ there is no difference in the inhibitory effect per individual of the two species.)

In interference competition, coefficients of less than unity simply imply that whatever form of attack is being used, it has a less severe inter- than intra-specific effect. In some cases, such as allelopathy, this seems exceedingly unlikely.

In exploitative competition, which is what we are largely concerned with here, the conditions α, $\beta < 1$ have been interpreted in terms of niche differentiation. If species compete for a common limiting resource, and their usage of that resource differs in only a single dimension (say size of food item), then the numerical values of the competition coefficients are closely associated with the degree of overlap between the RUFs on that dimension (as shown in Figure 1.4). If the RUFs are of a similar size and shape, as shown, competition coefficients must fall in the range 0–1, and we would also expect to find $\alpha \simeq \beta$. Since both of these expectations are often untrue in actual experiments, it is clear that this simplified way of looking at the competitive situation has its limitations; but this does not necessarily mean that the whole business of relating competition coefficients with RUFs is erroneous. Indeed, if exploitative competition is taking place, some such relationship must exist, and modellers of exploitative competition continue to assume such a relationship (e.g. Loeschcke, 1984).

As an aside, I should point out here that while some pairs of α and β values, including those yielding stability ($\alpha < K_1/K_2$, $\beta < K_2/K_1$), can be interpreted in terms of either exploitation or interference, others would seem to preclude an interpretation solely in exploitative terms. This is the case when the product $\alpha\beta > 1$. The reason for this is as follows. Consider the hypothetical situation of exploitative competition between two species with equal carrying capacities and utilizing an identical range of resources, but with one consuming twice the quantity of resource per unit time that the other consumes (perhaps because of larger body size). In this situation we would have $\alpha = 0.5$, $\beta = 2$, and $\alpha\beta = 1$. Other situations can be imagined where, within the general constraint of $\alpha\beta \leqslant 1$, different numerical values will be found. However, if we obtain, say, $\alpha = 2$, $\beta = 2$, it is hard to see how this could be explained without invoking at least an interference effect in one direction. There are examples of experiments where $\alpha\beta > 1$, but we should be hesitant to come to the conclusion that this necessarily indicates interference unless we know the confidence limits of the estimates of α and β.

In conclusion, the L–V model tells us that if two species are to coexist, certain inequalities must be satisfied. This has been interpreted by niche theorists as meaning that there must be some degree of separation of RUFs. Since we are dealing here with a strategic model, as discussed earlier, what is important is not whether $\alpha < K_1/K_2$ and $\beta < K_2/K_1$ are conditions for coexistence that can be precisely confirmed in practice; but rather whether the general association between RUF overlap and severity of competition, and the existence of some (probably variable) limit to RUF overlap compatible with coexistence, are phenomena that characterize the world of real organisms competing in real environments—both laboratory and natural.

Investigation of this question can take two forms: (a) demonstrations of competitive exclusion coupled with attempts to link this with lack of niche differentiation and (b) demonstrations of coexistence, coupled with attempts to link this with some, or a certain degree of, niche differentiation. The next two sections discuss the first approach, Chapter 5 the second. Chapter 5 also examines some alternative models of interspecific competition. The principles of competitive exclusion and limiting similarity, which are central in this area of research, are discussed in Section 5.6. For the moment I will content myself with making the elementary point that competitive exclusion is a *process* which definitely happens as the next section will show, whereas the competitive exclusion *principle* is an attempt to generalize about the conditions under which competitive exclusion (or coexistence) occurs. The principle is therefore related to, but is not the same as, the L–V model.

2.2 EXPERIMENTAL ILLUSTRATION

Numerous laboratory competition experiments have been conducted on populations of a variety of organisms, particularly yeasts, *Paramecium*, *Tribolium* and *Drosophila*, and most of these experiments have resulted in competitive exclusion. It would be futile to attempt an exhaustive review of this vast literature; instead, I will take a highly selective approach and discuss in detail only three sets of experiments—*Callosobruchus* (Bellows and Hassell, 1984), and two 'classic' experimental systems, namely *Paramecium* and *Drosophila*. There are some reasons for my selecting *Paramecium* and *Drosophila* in preference to other classic systems such as *Tribolium*. First, it makes sense to concentrate here on systems which have counterparts resulting in stable coexistence which will be discussed in Chapter 5. This rules out the extensive work of Park (1948, 1954, 1957) and his colleagues on *Tribolium*, since none of the *Tribolium* experiments resulted in stable coexistence. Second, it is long-term experiments in which the resource is periodically renewed that are maximally informative about natural populations. Thus Gause's (1932) experiments with yeasts, in which the resource was not renewed, will be omitted. Gause's experiments on *Paramecium*, on the other hand, were of the resource-renewal type, and included both exclusion and coexistence outcomes. Also, the experiments resulting in coexistence were published in Gause's (1935) book written in French. It is abundantly clear from the lack of, and often incorrect, information given about these experiments by those referring to them that many of the authors concerned were using 'second-hand' sources written in English. I have taken some trouble to give a detailed account of these 'French' experiments in Chapter 5, so it makes sense to consider their 'English' counterparts (Gause, 1934), which resulted in competitive exclusion, here. Finally, I have chosen to include the *Drosophila* system partly because of my own personal involvement with it, partly because of the interest of geneticists in *Drosophila*, and partly because this system has revealed more cases of stable coexistence (i.e. involving more species pairs) than any other laboratory system.

I will not be discussing Crombie's (1945, 1946) well-known competition experiments on granivorous insects in detail, but will make just three brief comments on them here. First, these experiments are rather confusing since Crombie attempts to explain the results of the first series in terms of classical resource partitioning, whereas he explains the results of the second in terms of predation and 'cannibalism'. Second, some of the interactions were not competition at all. Some calculations on his *Sitotroga–Oryzaephilus* experiments show the equilibrium population size of *Oryzaephilus* to be approximately 440 in both single and mixed culture, while that of *Sitotroga* is depressed when mixed—so the interaction appears to be an amensal one. Third, the attempted explanation of some coexistences in terms of resource partitioning is open to the same criticism I level at Gause (in Chapter

5)—namely that there is no quantitative evidence for the proposed form of partitioning.

2.2.1 *Paramecium*

The most extensive series of competition experiments on *Paramecium* was conducted by G. F. Gause and his co-workers in the 1930s. These experiments are reported in two books (Gause, 1934, 1935) and in several related papers (Gause, 1936, 1937; Gause *et al*., 1934; Gause and Witt, 1935). More recent experiments using *Paramecium* were conducted by Vandermeer (1969).

The majority of Gause's experiments involved pairwise combinations of the three species *P. aurelia*, *P. bursaria* and *P. caudatum*, and I will restrict the discussion to these experiments. Of these three species, *P. aurelia* and *P. caudatum* are similar ecologically, in that both feed, according to Gause, in the upper layer of the food-medium in the culture tubes. *P. bursaria*, on the other hand, feeds predominantly in the lower layers. This distinction could easily be represented in terms of one-dimensional RUFs; regrettably, however, Gause provides us with no such data. We will proceed with the uneasy assumption that this proposed niche difference between *P. bursaria* and the other two species is a real one. One obvious area for future research is to provide the missing quantitative data on this phenomenon.

The experiments resulting in competitive exclusion (Gause, 1934) involved *P. aurelia* and *P. caudatum*. With all of the types of culture medium that Gause employed, both of these species were able to establish stable monoculture populations. The pattern of growth up to the carrying capacity was approximately logistic in most cases. Which species has the higher carrying capacity depends on whether the data are plotted simply as numbers of individuals, or as 'volume'. Gause (1934) often employed the latter form of presentation, wherein the numbers are weighted to take into account the greater size of *P. caudatum* individuals (This 'volume' measure is similar to biomass, though of course expressed in different units.)

When mixed cultures of the two species were established, both species were depressed in numbers, indicating the occurrence of competition. However, while *P. aurelia* merely reached a lower equilibrium, *P. caudatum*, after an initial phase of growth, gradually declined in numbers for the rest of the experiment (see Figure 2.1).

It is unfortunate that Gause (1934) did not take these experiments through to their expected finishing point—i.e. the competitive exclusion of *P. caudatum*. As the results stand, we may acknowledge that they exhibit a clear *trend towards competitive exclusion*, but not exclusion itself. Many other long-term competition experiments, conducted by other workers, were also prematurely terminated. The problem here is that 'reversals of dominance' have often occurred in long-continued experiments. These have been

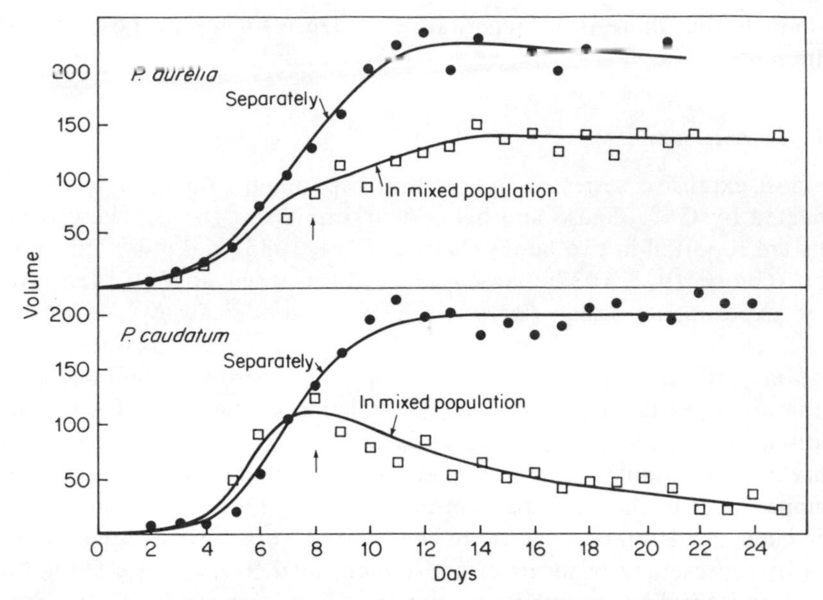

Figure 2.1 Trend towards competitive exclusion of *Paramecium caudatum* by *P. aurelia* (open squares), compared with growth to carrying capacity in monocultures (solid circles). From Gause (1934)

observed in *Drosophila* (Ayala, 1966, 1969) and in other dipterans (Pimentel *et al*., 1965), and a recent theoretical treatment of this phenomenon has been presented by Pease (1984). Whatever the reasons for such reversals, and there are probably several, their occurrence means that we cannot safely assume that the smooth downward trajectories of *P. caudatum* numbers (or volume) will continue until the populations concerned reach extinction. If they do so continue, we would expect that *P. aurelia* would also return to its monoculture *K*-value; but again the experiments were not continued for long enough to allow direct observation of this return.

With regard to the *direction* of competitive exclusion, all of Gause's (1934) long-term experiments showed that *P. aurelia* out-competed *P. caudatum*. This outcome was consistent both across different replicate populations on the same culture medium and across different media. A separate series of experiments in which the culture medium was not periodically renewed (Gause *et al*., 1934) suggested that there might be conditions under which *P. caudatum* would 'win'; but long-term experiments to confirm this appear not to have been done.

Since Gause's extensive work with *Paramecium*, it has been discovered that the 'species' *P. aurelia* in fact consists of several distinct groups or syngens, which are really separate species in their own right. It seems unlikely that this detracts in any way from the conclusions deriving from Gause's work. Presumably, Gause was using one particular syngen throughout; we do not of course know which one, but this is of little consequence in relation to the elucidation of the dynamics of competitive exclusion.

The most important question, in relation to the *Paramecium* experiments described above, is why competitive exclusion took place. Gause very strongly advocated a niche-based explanation of his results. He states (Gause, 1937) that 'the steady state of a mixed population consisting of two species occupying an identical "ecological niche" will be the pure population of one of them, the one better adapted for the particular set of conditions'. This is one version of the competitive exclusion principle, other versions of which will be given in Chapter 5.

The most severe objection to a niche-based interpretation of Gause's experiments is that we have no quantitative data on the degree of similarity/difference of the niches of *P. aurelia* and *P. caudatum* along either the dimension of postulated importance (depth of feeding) or any other. This is important, not only in relation to Gause's experiments, but as a general criticism of studies of this kind on a wide range of organisms. Basically, while studies revealing stable coexistence have sometimes attempted a quantification of proposed niche differences, experiments revealing competitive exclusion almost never involve attempts at quantifying the proposed lack of niche differences. As noted in Chapter 1, a sceptic could well argue that niche differences always exist, but will only be found where they are looked for, namely in experiments revealing coexistence. Until this problem is remedied, the interpretation of the results of competition experiments is problematic, and alternatives to niche-based explanations remain possible. In this context, Case and Gilpin (1974) have argued that Gause's experiments may have involved interference rather than (or as well as) exploitation. If so, then the reason why one species excludes (or coexists with) another cannot be phrased in terms of niches.

2.2.2 *Drosophila*

About two decades after Gause's experiments with *Paramecium*, pairs of *Drosophila* species became a popular system in which to study the dynamics of interspecific competition. Among the first detailed investigations was the work of Moore (1952a,b) on the pair of sibling species *D. melanogaster* and *D. simulans*. Many subsequent studies have been performed, including the well-known work of Ayala (1969, 1970, 1971), and some of the most recent work has emanated from my own laboratory (Arthur and Middlecote, 1984a,b; Arthur, 1986). Competition experiments with *Drosophila*, only a tiny fraction of which are referred to above, have involved a number of different species-pairs. (There is a wealth of species to choose from—currently more than 2000 known species in the genus *Drosophila*.) These experiments have been reviewed, though by no means exhaustively, by Barker (1983).

I will concentrate here on experiments with *D. melanogaster* and *D. simulans*. This pair of species is apparently never capable of stable coexistence, or at least if it is, the many long-term experiments that have been conducted have failed to find a set of conditions under which stable coexistence can be conclusively demonstrated. Exclusion appears to be the universal outcome of such experiments, though which direction it takes is variable; i.e. unlike Gause's (1934) *Paramecium* experiments, the same species is not always excluded.

The variable direction of competitive exclusion has a definite pattern, with some conditions favouring *D. melanogaster* and some *D. simulans*. When wild-type stocks of both species compete on a standard laboratory food-medium at 15 °C, it seems that there is a trend towards competitive exclusion of *D. melanogaster* (Moore, 1952a). This is one of the least well established outcomes because experiments run at such a low temperature take a very long time to complete, and have always been 'prematurely terminated' in the sense noted earlier. At 25 °C, the most frequently used temperature for these experiments, *D. melanogaster* is a clear winner, rapidly excluding *D. simulans*. However, if certain mutant stocks of *D. melanogaster* are employed, *D. simulans* wins despite the higher temperature. Introduction of a high concentration of alcohol into the culture medium favours *D. melanogaster* and can result in exclusion of *D. simulans* even by a rather drastically mutant stock of *D. melanogaster* (Arthur, 1980a).

These results, which are summarized in Table 2.1, nicely illustrate that the outcome of competition beween a particular pair of species depends on the interplay between a variety of genetic and ecological factors. It must be admitted that the genetic status of some of the *D. melanogaster* stocks used was distinctly unnatural; but it is most unlikely that the genetic differences that exist within and among natural populations do not also influence the outcome of competition.

Under most specified sets of genetic and environmental conditions, the direction of competitive exclusion is consistent between different replicate populations. However, if the net effect of a particular set of conditions is such that the performances of the two species are very close, the direction of

Table 2.1 Outcome of competition between *Drosophila melanogaster* and *D. simulans* under a range of conditions

Temperature (°C)	Genetic status		Ethanol[b]	Winning species
	melanogaster	*simulans*		
15	wild	wild	—	*simulans* (?)
25	wild	wild	—	*melanogaster*
25	mutant[a]	wild	—	*simulans*
25	mutant[a]	wild	+	*melanogaster*

[(a)]Mutant stocks with considerably reduced viability; [(b)]a plus in this column indicates ≥8% concentration.

competitive exclusion may depend on factors which are not controlled at the start of, or during, the experiment. Such factors include genetic sampling variation—i.e. genetic differences between the groups of founders used for different populations—and, in many but not all experiments, relative humidity. An example of this phenomenon is provided by the results of competition between *D. melanogaster* and *D. simulans* on medium containing 4 per cent ethanol, using stocks with which *D. simulans* wins consistently at zero per cent and *D. melanogaster* at 8 per cent. At the intermediate alcohol level, *D. melanogaster* outcompeted *D. simulans* in one population, while the opposite outcome was obtained in a further five populations (Arthur, 1980a). This phenomenon, referred to as *indeterminacy* in the direction of competitive exclusion, has been extensively studied in *Tribolium* (see Park, 1954, 1962). It is usually very difficult to determine which uncontrolled factor is responsible; and it seems likely that different instances of indeterminacy are caused in different ways.

An illustration of the more usual situation where competitive exclusion is consistent in direction is given in Figure 2.2. The mixed cultures whose fate is shown were started with either 20 *D. melanogaster* and 80 *D. simulans* or the converse. However, the data are presented not as straightforward numbers but as *species frequency*. This is simply the number of one species as a fraction of the total numbers. In the case of two species, the frequency of species 1 is $N_1/(N_1 + N_2)$ and can be given either on a percentage or probability scale. One obvious consequence of presenting data in this way is that the results for each mixed culture appear as a single trajectory rather than two. Straight-

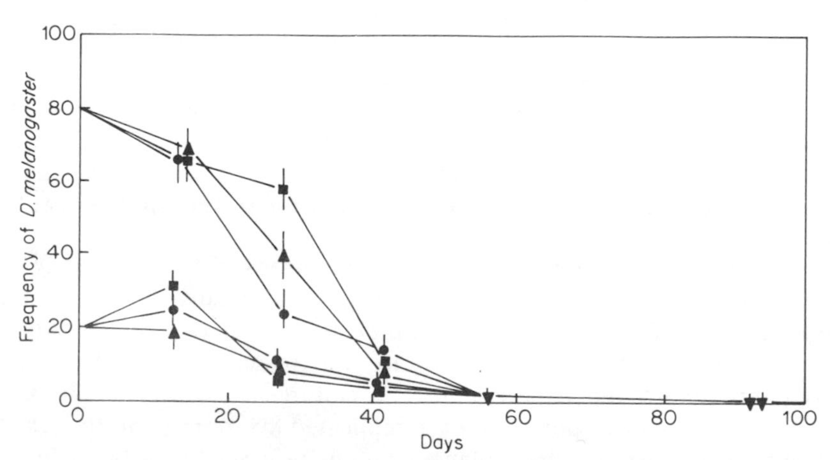

Figure 2.2 Competitive exclusion of *D. melanogaster* by *D. simulans* under the third set of conditions listed in Table 2.1. ●, ■, ▲—different replicate population cages. Bars indicate 95 per cent confidence limits. One generation ≃14 days at 25 °C. Reproduced by permission of Academic Press Inc. (London) from Arthur (1980) *Biol. J. Linn. Soc.*, **13**, 109–118

forward competitive exclusion, without reversals of dominance 'within' a replicate or indeterminacy between replicates, has a particular pattern when illustrated on a frequency-against-time graph, as shown in Figure 2.2.

One advantage of using frequencies is the greater degree of condensation of the data, and another is that the parallel between transient coexistence and transient polymorphism (or between stable coexistence and balanced polymorphism) is more obvious, as the state of a polymorphism is normally given in terms of the gene frequency, q. (For a two-allele polymorphism, $q_1 = n_1/2N$, where n_1 refers to the number of copies of allele 1 in the population: see Chapter 3.) A disadvantage of frequency data is that the same frequency can be arrived at in different numerical ways. This problem is discussed further in Chapter 4.

The experiment whose outcome is illustrated in Figure 2.2 was chosen from many similar experiments partly because it was continued all the way to competitive exclusion. Thus one of the two problems noted in relation to Gause's experiments was avoided. However, the other problem, namely lack of quantification of niche similarity, remains. Now this is not a problem if the aim is solely to demonstrate competitive exclusion. It becomes a problem, though, if the experiment producing exclusion is contrasted with another in which coexistence, allegedly due to niche differences, is observed. I will postpone discussion of such a contrast, and its associated problems, until Chapter 5.

2.2.3 *Callosobruchus*

A recent experimental study of competition between the beetles *Callosobruchus chinensis* and *C. maculatus* was carried out by Bellows and Hassell (1984). These authors adopted a particularly interesting approach which, in retrospect, it is surprising has not been adopted more often. They devised three alternative models of the competitive interaction between these two species, used short-term experiments with the beetles to produce estimates of the various parameters in the models, and then ran long-term, multigeneration experiments and compared the results with the models' predictions.

The long-term experiments involved mixed cultures of the two species of *Callosobruchus* maintained on a limited supply of 'beans' of the legume species *Vigna unguiculata*. As with the *Paramecium* and *Drosophila* experiments discussed in the previous sections, the resource was periodically renewed throughout the experiment. The cultures were started at three different combinations of densities, each replicated five times, and the results are shown, in the form of phase-plane diagrams (see Section 4.1), in Figure 2.3.

Several features of these results deserve mention. First, and perhaps most obvious, there is competitive exclusion in all cultures (and none were 'prematurely terminated'). Second, in fifteen out of sixteen cases *C. chinensis* was the

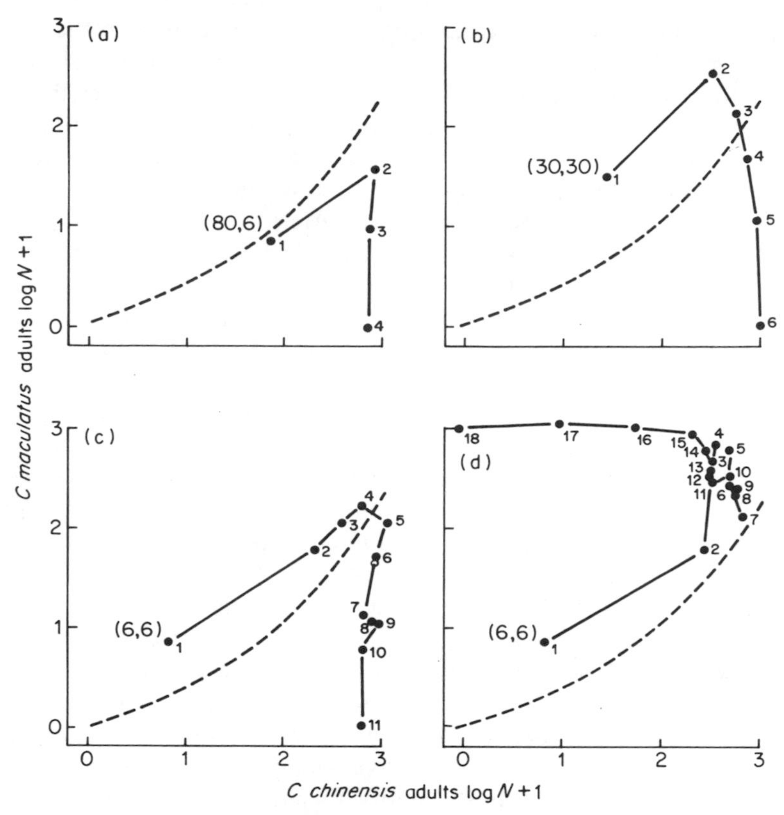

Figure 2.3 Competitive exclusion in *Callosobruchus*. Numbers in brackets indicate starting densities. Numbers beside the data-points indicate the time-sequence of the observations. (Consecutive observations are three weeks apart.) The dashed line divides the phase plane into an upper sector (where *C. maculatus* is the predicted 'winner' in the single age-class model) and a lower sector (where *C. chinensis* is predicted to win). (a), (b) Mean of five replicates. (c) Mean of four replicates. (d) Single replicate. Reproduced with permission from Bellows and Hassell (1984)

winning species. The single exception—one of the replicates started at a density of six of each species—resulted in exclusion of *C. chinensis*. As in other cases of indeterminacy in other systems it is difficult to pin down the cause of the departure of some replicates (in this case just one) from the predominant outcome. The pattern of system behaviour in the aberrant replicate here (bottom right diagram in Figure 2.3) suggests that a stable equilibrium was temporarily achieved but was then 'broken out of' somehow—e.g. by evolutionary change in the *C. maculatus* population. However, further work would be necessary to test such a hypothesis, and at any rate it still would not explain why that culture began to diverge from the others by heading for a stable equilibrium in the first place.

Comparison of the results of the long-term experiments with the predictions of the models reveals some interesting points. Two of the models were of the conventional, 'analytic' kind—one allowing for age-structure, the other not. Neither of these models were able to predict the correct outcome of the experiments. This is illustrated, for the single age-class model, in Figure 2.3 (see Figure caption for explanation). As Bellows and Hassell (1984) point out, this lack of predictive ability appears to result from the limitations of the single-generation experiments on which these models were based. Specifically, *C. maculatus* seems to be a superior competitor within a single generation, which is what the short-term experiments measured; but *C. chinensis* has a faster development, and this may enable it to begin utilizing each fresh batch of resource before *C. maculatus* in a long-term, renewed-resource experiment. The authors propose that this can override the apparent within-generation superiority of *C. maculatus*. This claim is supported by the prediction of a third, 'systems' model which incorporated an interspecific difference in development time. The systems model predicted that *C. chinensis* would always exclude *C. maculatus* in long-term experiments, and clearly the data are largely in accordance with this prediction.

My main criticism of this case study is the same as in the cases of *Paramecium* and *Drosophila*, discussed above. That is, we are given little information on patterns of resource utilization by the species concerned. If the competition is exploitative, then it seems likely that the two species of *Callosobruchus* use the 'beans' in a very similar way, since competitive exclusion is the result. However, as I have aleady pointed out in relation to the other case studies, it is unsatisfactory to have to *assume* lack of niche differentiation and there is a danger in general of making an erroneous link between niche differences and coexistence if differences are only looked for in cases where coexistence, rather than exclusion, is the result.

2.3 NATURAL POPULATIONS

There is probably not a single case in which we can conclusively attribute the disappearance of a species from, or its lack of successful invasion of, a natural environment to competition with another species. This is largely because we cannot, in contrast to the laboratory experiments discussed above, set up replicate monoculture populations to demonstrate that the disappearing (or non-appearing) population would have persisted (or invaded) in the absence of another, resident species. Nevertheless, since our interest is ultimately in what happens in nature, it is necessary to persevere with whatever inadequate data we have at our disposal on natural populations, and to attempt to weigh up the alternative merits of competitive and other hypotheses as explanations of species distributions in time or space. As usual in ecology, laboratory experiments give us clear conclusions whose relevance to nature may be debatable, while field studies can hardly fail to be relevant to nature, but are rarely conclusive. Given such a situation, it is unwise to neglect either approach.

Field data suggestive of competitive exclusion fall into two distinct categories: (a) data on the relative spatial distributions of two or more species at a particular time and (b) data on the invasion of an area by an immigrant species and the concomitant decline of a previously-resident species over a period of time. Williamson (1972) describes the latter situation as a *displacement*, to distinguish it from situations in which one species gives way to another in space, which he calls a *replacement*. There are two difficulties with using the term displacement in this way. First, the word itself implies an active effect of one species on the other, and hence prejudges the issue. Second, the most frequently invoked form of coevolution resulting from competition is displacement (of morphological characters and/or niches: see Brown and Wilson, 1956) so there is potential confusion through use of the same term for two different but related phenomena. It seems more sensible to use the term replacement for both of the situations discussed in this section, by qualifying it with 'temporal' or 'spatial'.

Of the two types of data, those on spatial replacement are the weaker as evidence of competitive exclusion, but even so some cases are highly suggestive. One of these concerns sand dune populations of the snails *Cepaea nemoralis* and *C. hortensis* on the East coast of Great Britain. Coming northwards up this coast from southern England to northern Scotland, we find that up as far as the Scottish village of St Cyrus (about half way between Dundee and Aberdeen), only *C. nemoralis* is found. After that, only *C. hortensis* occurs on the dunes. The lack of *C. nemoralis* in northern duneland habitats appears to be unrelated to the presence of *C. hortensis*. Rather, it is merely a reflection, in one particular habitat, of the general lack of *C. nemoralis* in northern Scotland. (The northern limit of *C. nemoralis*' geographical range runs through Scotland; that of *C. hortensis* is further north, and cuts through central Iceland.) In contrast, the lack of *C. hortensis* in more southerly dunes—with the exception of two small populations at St Cyrus itself and at Seaton Sluice in Northumberland (Arthur, 1978, 1982b)—does seem most readily explained in terms of competitive exclusion by *C. nemoralis*.

Alternative explanations cannot be dismissed, but it is difficult to construct one that does not stretch our acceptance of coincidence a little too far. For example, we could postulate that *C. hortensis* can only inhabit duneland habitats at relatively low temperatures, and that this accounts for its absence in such habitats in England and southern Scotland, despite the occurrence of many inland populations in these areas. For this hypothesis to hold, we have to believe that the upper temperature limit for *C. hortensis* in dunes is reached, as we progress southwards, at exactly the same spot (to within a few kilometres) as the northern limit of *C. nemoralis*' geographical range. We cannot rule out such a coincidence, but we ought not to be too ready to accept it either.

We might tentatively accept the competition hypothesis as the more plausible of the two. Its plausibility is further increased by evidence of considerable overlap in foods utilized in natural mixed populations of *Cepaea*

(Carter *et al*., 1979), evidence of interference between the species in laboratory cultures (Cameron and Carter, 1979), evidence that some sort of interaction between the two species does take place in field population cages (Tilling, 1985a,b) and recent evidence of a *temporal* replacement of *C. nemoralis* by *C. hortensis* (Cowie and Jones, 1987). However, there are still many questions unanswered. One is why competition should prevent the existence of mixed-species colonies on sand dunes but not in many other habitat types. We could speculate about a simpler environment enforcing greater niche overlap—but no quantitative data exist to confirm such a possibility. Another question is whether the situation is one of competition or amensalism, since we have no evidence of any exclusions of *C. nemoralis* by *C. hortensis* in sand dunes. (Cowie and Jones's study, mentioned above, was conducted in the Marlborough Downs.) Amensalism can result in exclusion of the inhibited species just as competition can result in exclusion of the competitively inferior one, yet there is not even any term (e.g. amensal exclusion) for such a process. This is one of the many cases in which the terminology relating to ecological interactions among closely-related species involves an inbuilt assumption that such interactions are of a (−, −) nature. Finally, even if there is a competitive interaction between the two *Cepaea* species, it need not necessarily be exploitative competition; and if it is exploitative, there arises the question of what is the limiting resource. Two obvious candidates are food and the supply of suitable microsites, but we have no hard evidence to confirm that either of these really is limiting. In conclusion, even if we accept that one species has competitively excluded another in this case, we do not know how it has done so.

Turning to temporal replacement of one species by another, one of the most celebrated cases of this happening in a natural environment is that of the replacement of the red squirrel *Sciurus vulgaris* by the grey squirrel *S. carolinensis* in much of England and Wales. This replacement has been documented by several authors, notably Lloyd (1962, 1983), Middleton (1931, 1935) and Shorten (1953, 1954). The basic picture is that during this century the immigrant grey squirrel has spread through most of southern and central England and Wales, while the native red squirrel has disappeared from these same areas. As in the case of the spatial replacement in *Cepaea*, this can be accounted for either in terms of competition or coincidence.

A detailed study of a restricted area (Norfolk) was carried out by Reynolds (1985). The grey squirrel has invaded Norfolk relatively recently—over the last 20 years or so—which should make the data more reliable than that from areas in which the replacement occurred in earlier decades. Reynolds provided detailed maps, based on 5 km squares, for each of 21 consecutive years during the replacement, from 1960 to 1981. Those for 1960, 1970 and 1980 are reproduced in Figure 2.4. It can be seen that the grey squirrel invaded the county from the south-west, subsequently spreading through most of it. The red started off occupying most of the county, but declined dramatically,

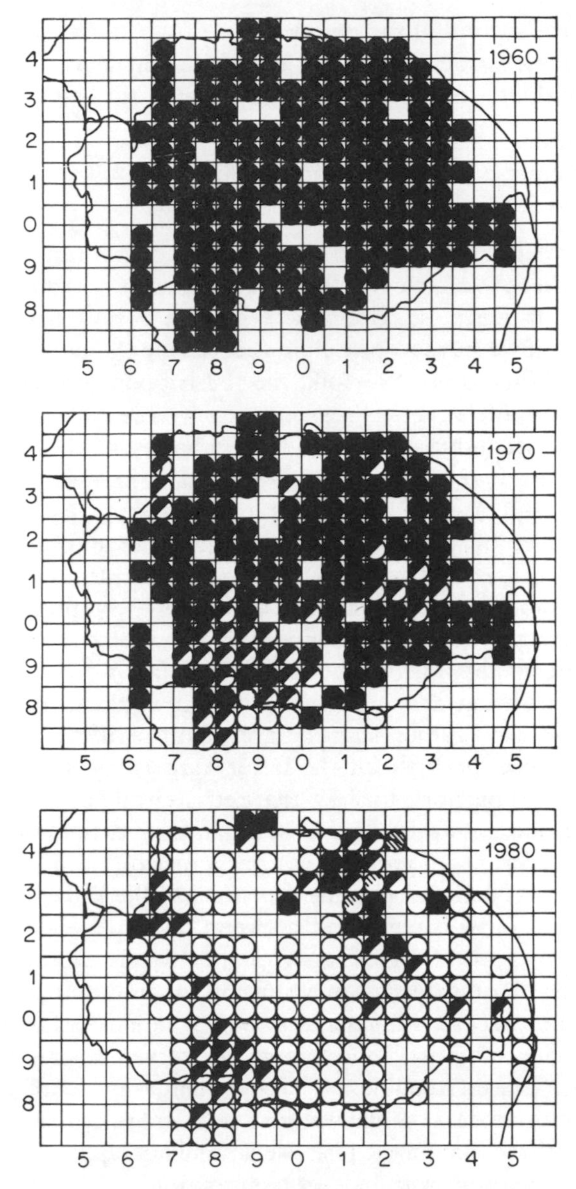

Figure 2.4 Replacement of the red squirrel by the grey in Norfolk, England over a 20-year period. ● Indicates presence of red squirrels. ○ Indicates presence of grey squirrels. ◐ Indicates presence of both red and grey. Cross-hatching indicates symptoms of disease in red squirrel populations. Numbers relate to National Grid. Reproduced by permission of the British Ecological Society from Reynolds (1985) *J. Anim. Ecol.*, **54**, 149–162

particularly in the second of the two decades of the study, and occupied only scattered localities, particularly in the north-east, in 1980.

Reynolds considers three hypotheses to account for this replacement:

1. The replacement represents competitive exclusion.
2. The replacement is a result of habitat changes.
3. The replacement is due to disease.

How do these relate to the distinction between competition and coincidence that I made earlier? Well, clearly the first hypothesis is the competitive one. The third is coincidence, so long as the disease is not carried by the grey squirrel. That is, if we have to postulate that during the two decades in which the grey squirrel spread into Norfolk, the red happened to contract a severe disease for some quite separate reason, then we are of course appealing to coincidence. The second hypothesis in fact embodies neither competition nor coincidence. Given that habitat changes are occurring—partly as a result of human activities—it is quite possible that these changes would make the continued existence of a resident species impossible and, at the same time, invasion by a new species possible. This is not coincidence in the sense that the two events stem from the same cause. Nor is it competition—unless the habitat changes act by switching competitive superiority.

Reynolds rejects the idea that habitat changes were important in the replacement, since what changes have occurred would seem, if anything, to favour the red squirrel rather than the grey. Left with the choice of competition or coincidence, he opts for the latter, largely on the basis of a single statistical test which purports to show that red squirrel populations, as defined by 5 km squares, do not have an increased chance of extinction when the grey is present. However, this approach can be criticized both on the basis that 5 km squares and functional populations are not the same, and also on the basis that the test just distinguished between presence and absence of grey squirrels—so ten or 10 000 grey squirrels per square would be treated the same. (Reynolds himself does draw attention to this latter problem.)

It seems to me that the willingness to accept coincidental explanations of phenomena such as temporal replacements is an over-reaction to the earlier, and now much criticized, readiness to accept competitive explanations despite the lack of conclusive data. While we need to be more critical of unsupported claims of competition than in the past, we should not equate lack of watertight evidence for competition with lack of competition.

2.4 EXCLUSION OF THE LESS FIT SPECIES

The end-result of the deterministic trend towards competitive replacement of one species by another is the actual 'exclusion event' itself—that is, the reduction from some small but finite population size to zero in the weaker competitor. This requires no special explanation, as it is part of the overall process and will happen inevitably when the population has declined to a very low level. It may be, though, that in practice there is a special explanation, and indeed a unique one in every situation, in that it will often be a stochastic event unrelated to competition that eliminates the last handful of survivors. Random elimination of small groups within a population is something that goes on all the time, particularly in natural populations, but its effects are insignificant when the population size is large. When numbers are very low, the effect of such a stochastic event can be catastrophic. However, it is still the deterministic process that is the interesting one. In an ideal world without random events, it would be competition itself that removed the last few individuals of the weaker species; and in the real world it is competition that reduces the population to a level at which it is at the mercy of stochasticity.

In a multi-species context, an additional aspect of exclusion becomes important. We need, in such situations, to enquire about whether the species that disappear from the community as a result of competitive exclusion are non-randomly distributed in some sense. One possibility here is that exclusions occur in such a way as to leave each resource spectrum populated by evenly-spaced RUFs. If so, competition is an important agent in the determination of community structure. The debate about whether this view of natural communities is correct will be discussed in Chapter 8.

Chapter 3

Directional Selection

3.1 BASIC THEORY

The basic idea of directional selection dates, of course, back to Darwin (1859). However, the basic *theory*, in mathematical form, had to wait until the present century. Much of selection theory was laid down by the 'great triumvirate' of population geneticists Fisher, Haldane and Wright in the 1930s (see e.g. Fisher, 1930). This theory was, thus, being developed at around the same time that Lotka and Volterra were formulating the first theory of interspecific competition and Elton was propounding the advantages of viewing natural communities in terms of the biotic niche (see previous chapters). Clearly, the 1920s and 1930s were a creative period in the development of population biology.

I will present, below, the results of some of the most elementary parts of selection theory relating to the subject of this chapter. From a theoretical viewpoint, directional selection is much simpler than either balancing selection or situations in which selection is absent and hence the fate of alleles is dependent on genetic drift. As with the treatment of the L–V model in the previous chapter, the theoretical discussion will be kept fairly brief, as this material is available in a variety of texts, though this time in population genetical, as opposed to ecological, ones (see e.g. Cook, 1971; Hedrick, 1985).

Sticking to the philosophy of focusing on the simplest situations, I will examine a two-allele polymorphism with genotypes A_1A_1, A_1A_2 and A_2A_2. It is necessary to distinguish different dominance conditions between the alleles, because the rate of spread of one allele at the expense of the other is affected by this. I will deal only with complete dominance of the favoured allele,

Table 3.1 Three possible dominance–fitness relationships

Situation	Genotypic fitnesses A_1A_1	A_1A_2	A_2A_2
Selectively favoured allele (A_2) is recessive	1 − s	1 − s	1
Selectively favoured allele (A_1) is dominant	1	1	1 − s
No dominance; selection in favour of A_1 allele	1	$1 - \frac{1}{2}s$	1 − s

complete dominance of the allele being selected against and no dominance (see Table 3.1). This leaves out partial dominance, which is actually a whole range of intergrading possibilities rather than just one (and therefore difficult to model) and what is sometimes confusingly called 'overdominance'—i.e. heterozygous advantage—as this takes us into the realm of balancing selection, which will be dealt with in Chapter 6.

Given that the gene frequency is q (see Section 2.2.2) and the selective coefficient acting against a genotype is s, then for the first case given in Table 3.1 the change in frequency over a single generation can be shown to be:

$$\Delta q = \frac{+sq^2(1 - q)}{1 - s(1 - q^2)} \tag{3.1}$$

Here and in the two subsequent cases, q represents the frequency of the A_2 allele. Now since s and q are bounded within a scale of 0 to 1, Δq in the above equation is always positive, as we would expect. The equation's usefulness is not in confirming that a favoured allele will spread to fixation, which is hardly in doubt, but rather in quantifying its rate of spread which, unlike the direction, does alter with the starting value of q. Figure 3.1 shows the overall pattern of spread to fixation of a favoured allele from a starting point of 0.1. This is the continuous equivalent of aplying equation 3.1 on 24 successive occasions in a situation where $s = 0.5$.

Equations 3.2 and 3.3 below quantify the rate of spread of the favoured allele (or actually, in the form given, the rate of decline of the one selected against!) for the second and third cases of Table 3.1, respectively. Thus for complete dominance of the selectively favoured allele A_1, we have

$$\Delta q = \frac{-sq^2(1 - q)}{1 - sq^2} \tag{3.2}$$

while for situations in which dominance is absent,

$$\Delta q = \frac{-\frac{1}{2}sq(1 - q)}{1 - sq} \tag{3.3}$$

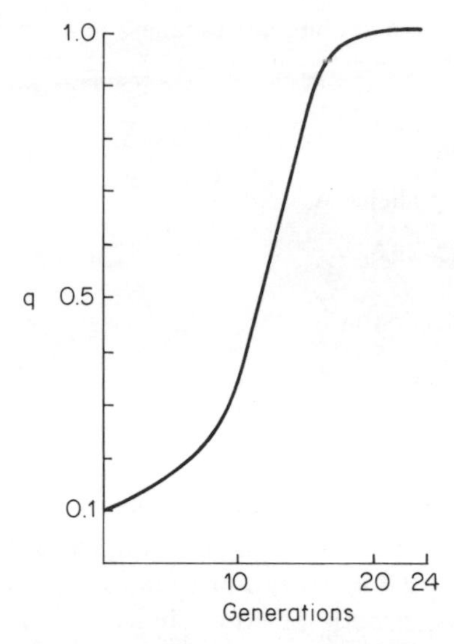

Figure 3.1 Pattern of changing gene frequency (q) under directional selection when $s = 0.5$ and $q_0 = 0.1$. From Arthur (1984)

As in the case of equation 3.1, these predict a unidirectional trend towards fixation. Thus all cases of directional selection lack a stable equilibrium in gene frequency, apart from the 'trivial equilibrium', to use the mathematician's label, at 0 or 1. Varying the dominance relationships alters the precise rate and pattern of spread, but not the stability characteristics of the system.

This last statement is only true if we exclude from consideration 'overdominance'. I would like to pause to consider this phrase, because it is a potential source of confusion, as noted above. In general, dominance of any kind refers to the interaction between alleles A_1 and A_2 in producing the value of some unspecified phenotypic character in the heterozygote. The character may be an 'ordinary' one such as pigmentation, which is the usual context in which we meet dominance. For example, if A_1A_1 is black and A_2A_2 white, a black heterozygote indicates dominance of the A_1 allele. If A_1A_1 is grey and A_2A_2 white, then a black heterozygote is properly referred to as overdominance. However, the concept of dominance is sometimes used in relation to a rather special phenotypic characteristic—fitness. This, indeed, is the way in which it is used in the context of Table 3.1. Where the heterozygote ranks highest of the three genotypes on a scale of fitness, the situation has conventionally been described as heterozygous advantage. 'Overdominance' used without qualification to mean the same thing is confusing; and 'overdominance in fitness', while correct, is a cumbersome phrase. It seems preferable to use the self-explanatory term heterozygous advantage to avoid this confusion.

The equations given above have some similarities to, and some differences from, the L–V model of interspecific competition. They are similar in both being strategic models; and in not specifying how the system works biologically. That is, just as in the case of a competitive exclusion occurring under L–V population dynamics, a fixation occurring under any of the three above equations may occur for many different reasons. The population may be limited by a resource or a predator. In cases of resource limitation, the 'winning genotype' need not be superior in acquiring the resource—it just has to be superior overall, i.e. fitter.

There are, however, some important differences between the genetic and ecological models (as well as some trivial ones, such as one being a difference equation, the other a differential). Interestingly, the L–V model together with other population ecology models is based on *numbers*, while almost all the models of population genetics are based on *frequencies*. As pointed out in Section 2.2.2, it is quite possible to have ecological models based on species frequencies. Also, it would in theory be possible to have genetic models based on pure numbers (of alleles, genotypes or phenotypes). Yet both these possible classes of model are rare or nonexistent. Why?

I suspect that this is a very loaded question, probing, as it does, into the basic philosophy of the disciplines of population genetics and population ecology. The former is concerned, always, with the success of one type of organism *relative* to another. Fitness, as normally defined, is a purely relative concept. On the other hand, much of population ecology is concerned with the dynamics of single species primarily and their interaction with other species only secondarily. Even in a very comparative branch of population ecology, such as competition theory, many researchers still stick to numbers, and the reason for this is not entirely clear. Whatever the reason, the avoidance of frequencies by many workers in the competition area has served to increase the gulf between their studies and those of population geneticists. It is in order to decrease this gulf that I often employ species frequencies in this book. (The opposite approach of converting genetical information into numbers is rarely possible, since gene-frequencies are usually based on a sample from a population.) The issue of choice of variables in ecology and population biology is discussed in more general terms by Maynard Smith (1974, Chapter 1); I say a little more on the numbers versus frequency issue in Chapter 4.

3.2 EXPERIMENTAL ILLUSTRATION

Demonstration of directional selection in the laboratory is extraordinarily easy. All that has to be done is to set up a polymorphic population of *Drosophila melanogaster* with some specified initial frequency, often 0.5, of a mutant allele; to let the population 'tick over' for a few generations; and to monitor the gene frequency at selected intervals through the experimental period. The selection coefficients acting against many commonly-used mutant

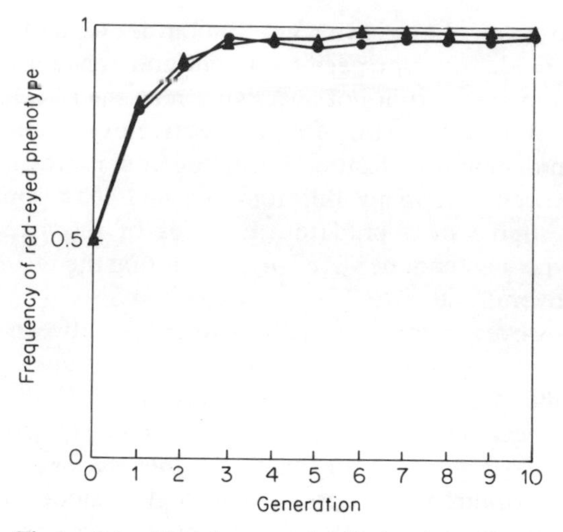

Figure 3.2 Trend toward elimination of the white-eyed phenotype and its replacement with the (red-eyed) wild-type in *Drosophila melanogaster*. (The two trajectories represent two replicate populations)

alleles (*vestigial*, *white*, etc.) are so great that selection is usually apparent in a single generation. Fixation, of course, takes longer, but even that may be observed within a handful of generations using mutations such as those mentioned above. The results of one such experiment, carried out in my own laboratory, are given in Figure 3.2. It should be noted that this experiment ended before the *w* allele was completely lost from the population—i.e. it was prematurely terminated, in the same way that many experiments showing competitive exclusion are. However, the 'reversals of dominance' which sometimes occur in competition experiments are never encountered in experiments on directional selection involving major mutants such as *white* so there is little reason to continue one's experiments to the bitter end.

3.3. NATURAL POPULATIONS

The extreme ease of demonstrating directional selection in the laboratory is almost matched by the extreme difficulty of demonstrating it in nature. Markedly detrimental mutations never achieve an appreciable frequency in natural populations. So we cannot observe selection acting against them. Of course, we presume that the reason such mutants do not reach appreciable frequencies is precisely because selection prevents them from doing so; but there is rarely any meaningful data on this 'selective prevention' for obvious reasons.

What would be of much more interest from an evolutionary viewpoint would be the spread through a population, from an initially low frequency on mutational input, of a new and selectively advantageous allele. Unfortunately, while this process may well be common (though very slow) in the 'invisible' alleles that contribute to continuous variation in characters like body size, it is very rare in the case of alleles causing discrete, individually-observable differences. Indeed, whether such alleles contribute to evolution at all is contested by some neo-Darwinists. (For details of this controversy, see Arthur, 1984; and Chapter 9.) On the rare occasions where 'major' polymorphisms are found in natural populations, they usually seem to be stable rather than transient. Examples include inversion polymorphism in *Drosophila* (Dobzhansky, 1970, Chapter 5), pigmentation polymorphism in *Cepaea* (reviewed by Jones *et al*., 1977; Clarke *et al*., 1978) and the sickle cell anaemia polymorphism in man (Allison, 1955).

It is because of the sparsity of cases in which a new allele producing a clearly defined phenotype has been observed to spread to (or near to) fixation in a natural population, that most accounts of this process, including the present one, are forced to use the example of industrial melanism in the moth *Biston betularia*. This is not, of course, to say that *B. betularia* is the only species in which we have clear evidence of directional selection—but there is still nothing to surpass it in terms of how much we understand about the nature of the process.

The extensive literature on industrial melanism in *B. betularia* (and other species) has been reviewed by Bishop and Cook (1980), who note that the genetic response to industrialization has been different in different places. The most detailed studies have been based on the replacement of the pale, speckled *typica* phenotype by the melanic *carbonaria* phenotype in the Liverpool/Manchester conurbation. Although data from this early stage of the replacement are very limited, it appears that the population of *B. betularia* in the early 1800s was, like other conspecific populations, exclusively or very predominantly *typica*. (The first recorded discovery of *carbonaria* was in 1848; see Bishop *et al*., 1978.) The frequency of *carbonaria* rose, however, from an initially very low value to, at its peak, more than 90 per cent. Both the genetic basis of, and the kind of selection on, melanic *B. betularia* are fairly straightforward. The *carbonaria* phenotype is produced by a dominant allele at a single locus. The main selective agent causing its spread in urban areas is known to be predation by birds, which eat disproportionately high numbers of *typica* resting on the blackened, lichenless bark of trees in industrial areas. (In rural areas the frequency of *carbonaria* is usually kept low by the same selective agent operating 'in reverse', since on pale, lichen-covered bark the *typica* is the better camouflaged phenotype.)

The basic picture of the replacement of *typica* by *carbonaria* in Liverpool/Manchester is a simple one. We presume that the system behaved roughly in accordance with the pattern of allele-replacement shown in Figure 3.1, though the timescale was longer as the selective coefficient was lower than the rather unrealistic 0.5 upon which that figure is based. (Also, Figure 3.1 shows the spread of a recessive allele whereas *carbonaria* is dominant.)

There are several complications to this basic picture, which are as follows. (A fuller account is given by Bishop and Cook, 1980.)

1. In different urban areas, melanic phenotypes with different genetic bases have increased in frequency instead of, or along with, *carbonaria*. These include *insularia* in Southern England and *swettaria* in the United States.
2. It is conceivable that *carbonaria* has some other selective advantage in addition to its superior camouflage on urban backgrounds; but conclusive evidence for this is lacking.
3. The frequency of *carbonaria* is now declining; this is attributed to pollution control measures having created a new kind of urban environment in which directional selection once again favours *typica* (Clarke *et al*., 1985; see Figure 3.3).
4. Before *carbonaria* began this decline, it had in several urban areas approached, but never actually reached, fixation. There are several possible reasons for this, including:
   a. Lack of time.
   b. Migration between neighbouring urban and rural populations.
   c. The action of some unknown form of balancing selection superimposed on the known directional selection.

While these, and perhaps other, complications must be acknowledged, industrial melanism in *Biston* remains the best example to date of directional selection operating on a phenotype of known genetic basis and known selective advantage in a natural population. Other good examples of this process tend also to be associated with human modification of the environment, largely because such modifications are often drastic, thus creating stronger selection pressures than found elsewhere. Other well-established selective responses to man's activities include evolution of heavy metal tolerance in plants (Antonovics *et al*., 1971; Bradshaw, 1984), and evolution of insecticide resistance (Sawicki and Denholm, 1984).

3.4 EXTINCTION OF THE LESS FIT VARIANT

In cases where the allele that is being eliminated is dominant, and in cases where there is no dominance, the final disappearance of that allele is no more complicated than the competitive exclusion of the weaker of two competing species (see Section 2.4). A complication arises, however, if the allele being selected against is recessive. When selection has reduced the frequency of an allele to a very low value, almost all copies of the allele are carried in heterozygous form. If we imagine an allele which is only very slightly deleterious, the genotypic frequencies will be approximately in Hardy–Weinberg equilibrium. If, then, the frequency of a deleterious allele is q, it can easily be shown that the proportion of copies of the deleterious allele carried in homozygous condition is also q—that is, this proportion is numerically equal to the gene frequency. So if, for example, $q = 0.01$, then only one out of every 100 copies of the allele concerned has its effect manifested.

Since, in this situation, heterozygotes display the phenotype of the favoured

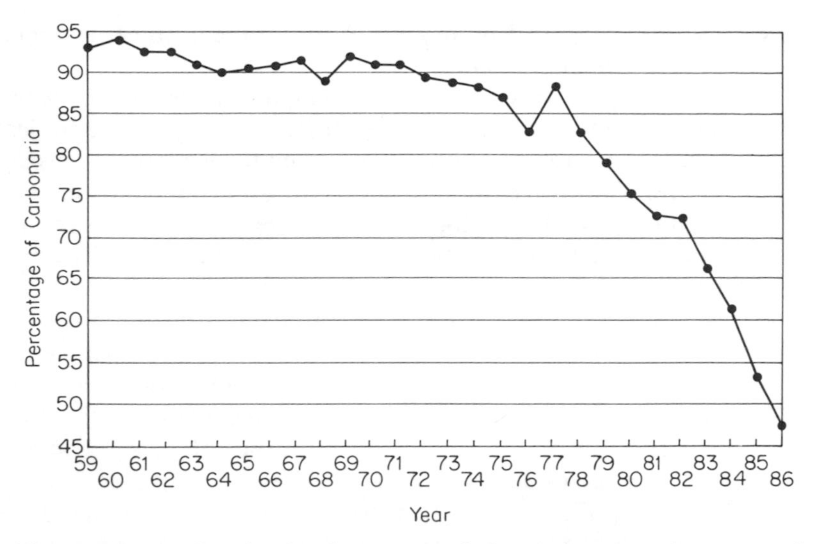

Figure 3.3 Decline in the frequency of the *carbonaria* phenotype of *Biston betularia* in an English population over the last 20–30 years. Reproduced by permission of Academic Press Inc. (London) from Clarke *et al.* (1985) *Biol. J. Linn. Soc.*, **26**, 189–199. (Supplementary data for 1985 and 1986 kindly provided by C. A. Clarke)

allele, they are not selected against except in as much as they will have some 'homozygous-unfit' progeny if they mate with another heterozygote. So what happens is that the more selection has achieved, in terms of pushing the frequency of the deleterious allele down towards zero, the less effective future selection is at causing further decline. Ultimately, then, we would expect the final elimination of the deleterious allele to be achieved by a stochastic event—i.e. by genetic drift. Indeed, even when the disappearing allele is not recessive, this may be its fate—just as stochastic processes may well be responsible for the final step in the competitive exclusion of a species (see Section 2.4).

3.5 HITCH-HIKING AND EPISTASIS

So far, this chapter has taken what might be described as a classical population genetics approach. That is, we have focused on a single locus, and looked at the selective replacement of one allele by another at that locus with almost no reference to the rest of the genome. When such an approach is adopted, the analogy between the elimination of a deleterious allele and the exclusion of a competitively inferior species is clearly seen. However, while it may be reasonable to consider the fate of a guild of two competing species in isolation from the rest of the community in which those species are embedded, it is almost certainly not reasonable to consider a single locus in isolation from the rest of the genome. (Counterviews to both of the above can be found. That is, there are those who propose a more atomic view of the genome, and a more holistic view of species interactions.)

Since the main purpose of this chapter is to emphasize the analogy between directional selection and interspecific competition, I do not intend to deal in much detail with the complications that arise when the genome is considered as a whole. Nevertheless, these must at least be mentioned and briefly described, for otherwise the analogy might be seen to be stronger than it actually is; and it could be argued that I have distorted the facts by omitting any mention of those genetic processes which have no ecological counterpart.

The two main processes that need to be considered here are hitch-hiking and epistasis, and I will deal with them in turn. The term hitch-hiking was first used in a genetic context by Kojima and Schaffer (1967). Basically, what it refers to is a situation where one locus (say A) is subjected to selection for one particular allele (say A_1) and the increasing frequency of A_1 is accompanied by increasing frequency of an allele B_1 at a second locus B simply because there is linkage disequilibrium between the two loci. That is, B_1 increases in frequency not because it has any selective advantage, but merely because stretches of chromosome bearing A_1 tend also to bear a copy of B_1.

The importance of hitch-hiking in determining the allele-frequency of a neutral or nearly-neutral polymorphism at the B-locus depends both on the strength of selection at the A-locus and on the tightness of the linkage between A and B (as well as on the degree of initial disequilibrium between A_1/A_2 and B_1/B_2). These dependencies arise because of the continual tendency of recombination to break down disequilibrium between loci and so to randomize the alleles at one locus in relation to those at another. Clearly, the more tightly linked are the loci, the slower will the disequilibrium decay. Given a slow rate of decay, strong selection on allele A_1 leading to rapid fixation of that allele could conceivably fix allele B_1 also; whereas if selection is weak, even a slow decay of the disequilibrium may be sufficient to prevent selection on A_1 determining the fate of B_1. Thus a system such as melanism in *Biston* which involves particularly strong selection is the most likely kind to exhibit hitch-hiking effects. No such effects have yet been documented in *Biston*, and indeed we do not really know enough about its genome to look for them, but there are some apparent examples of hitch-hiking in other systems (see e.g. Pinsker, 1981).

The fates of alleles at two different loci may be intertwined either because of linkage (as in hitch-hiking) or because the loci interact in determining a phenotypic character and hence fitness. This latter process, which can involve genes which are completly unlinked, is known as epistasis. To illustrate this, let us suppose that we have two unlinked loci C and D which affect a phenotypic character whose value is linearly related to fitness. Three possible combinations of character-values for the genotypes C_1C_1, C_2C_2, D_1D_1 and D_2D_2 are given in Table 3.2. (The table omits heterozygous genotypes so as to avoid having to complicate explanation of epistasis between loci by entangling it with issues of dominance 'within' loci.)

Table 3.2 Illustration of the meaning of epistasis
Top: No epistatic interaction—additive combination of loci.
Centre and bottom: Epistasis—effects non-additive.
(Numbers represent phenotypic character values)

	C_1C_1	C_2C_2
D_1D_1	5	10
D_2D_2	20	25
	C_1C_1	C_2C_2
D_1D_1	5	10
D_2D_2	20	20
	C_1C_1	C_2C_2
D_1D_1	5	10
D_2D_2	20	30

The first scenario depicted in Table 3.2 is a lack of epistasis: the other two involve epistatic effects of different kinds. Table 3.2 makes it clear that epistasis between two loci is just a particular example of what a statistician refers to as a departure from additivity.

Finally, I introduced the complications of hitch-hiking and epistasis as genetic processes with 'no ecological counterpart'. It may be worth pausing to consider whether this is true. In the case of hitch-hiking I stand by my original statement. Since this process depends on physical attachment of genes on chromosomes, a phenomenon for which there is no ecological analogue, it would seem impossible to have any process equivalent to hitch-hiking in interspecific competition. However, if we turn to epistasis there may conceivably be an ecological counterpart. To set the scene, let us note that multiple alleles at a single locus correspond to a guild of many competing species; and different loci correspond to different guilds. Thus 'ecological epistasis' would occur if the relative competitive abilities of two species in one guild were altered by the presence/absence or relative abundance of species in another guild. Indeed, if C and D are guilds, combinations like C_1C_1 correspond to species and the numbers represent competitive abilities in some way, then Table 3.2 suffices to represent 'ecological epistasis'. Whether or not such a process is important is disputed among ecologists. This is essentially the same dispute to which Maynard Smith (1974) was referring when he posed the question: 'Does the extent to which actual ecosystems show properties of persistence or stability depend on the fact that the pairwise interactions between species would likewise, in isolation, lead to stability and persistence?' Like Maynard Smith I note that this question remains unresolved, though, also like Maynard Smith, I suspect that a 'yes' answer is nearer to the truth than a ‘no’. To put this another way—I do not believe that an ecological community has the same degree of cohesion as a genome, and I suspect that the neglect of ‘epistatic’ interactions in ecology will do less damage than a comparable neglect would do (and has done) to evolutionary genetics. No doubt other students of these processes will have diverse opinions on this unresolved matter.

PART II

STABILITY IN COMPETITION AND POLYMORPHISM

This part of the book develops the general theme of niche differences as a cause of stable coexistence of competing organisms, whether different genotypes of the same species or different species within a guild. The ease with which niche differences can permit stability in theory is contrasted with the difficulties of actually demonstrating that they do so in any particular case-study. Most practical studies that have been undertaken fail to exclude alternative explanations for observed states of stable coexistence or polymorphism. Some suggestions are made as to how future studies might produce more conclusive results. These suggestions are developed, in the context of interspecific competition, into an 'ideal experiment' which I hope will serve as a basis for discussion and refinement.

The idea that stable polymorphism may arise through niche differences has not given rise to any general principle—possibly because this stabilizing mechanism is seen as a relatively rare mechanism of polymorphic stability compared, for example, to heterozygous advantage and rare type mating advantage. However, the idea that niche differences can and frequently do allow stable coexistence of competing species is associated with a general principle or, more correctly, two interrelated such principles—those of competitive exclusion and limiting similarity. These principles have been much maligned, largely as a result of some particularly unhelpful statements of them which render them simply untestable assertions. While I would support those who dismiss the untestable versions of these principles, I do not think that we should abandon all attempts to generalize about the conditions required for stable coexistence of competing species in nature. I thus develop a 'mechanistic' statement of the competitive exclusion principle, of which there are two possible versions—'strong' and 'weak'—both of which are eminently testable. I hazard a guess that this mechanistic principle—in one of its two forms—is an accurate generalization about the natural world.

Chapter 4

Stability and the Niche

4.1 STABILITY AND EQUILIBRIA

It used to be possible for ecologists to agonize over the meanings of 'equilibrium' and 'stability'. Luckily, this is no longer so, at least in relatively restricted situations such as a pair of competing species or the populations of a predator and its prey. There are now generally-accepted usages of these and related terms in such situations and it is thus possible simply to inform (or remind) the reader of these usages, which I will do in the present section, and then to move on to the more biologically interesting topic of what causes stability. As will become clear shortly, the various stability concepts that I will describe in relation to interspecific competition are applicable also to polymorphism, though polymorphic systems have the added complexity that there are more variables to choose from for describing the system, as we can focus attention on the frequency of the alleles themselves or alternatively on the frequencies of genotypes or phenotypes.

What follows is a concise description of basic stability concepts in the context of interspecific competition. More general discussions are given by May (1974b) and Maynard Smith (1974), and an equivalent description of stability concepts in relation to predator–prey systems has recently been provided by Taylor (1984, Chapter 3).

Stability concepts can be pictured in one of two manners, which are distinguished by the way in which time is represented. One approach is to construct a picture in which time forms the x-axis and some other variable, in terms of which the state of the system is described, forms the y-axis. The alternative is to construct a diagram whose axes are both (or all, if it is more

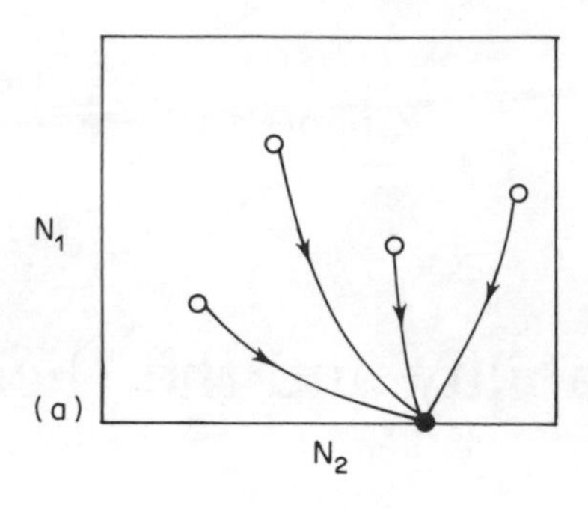

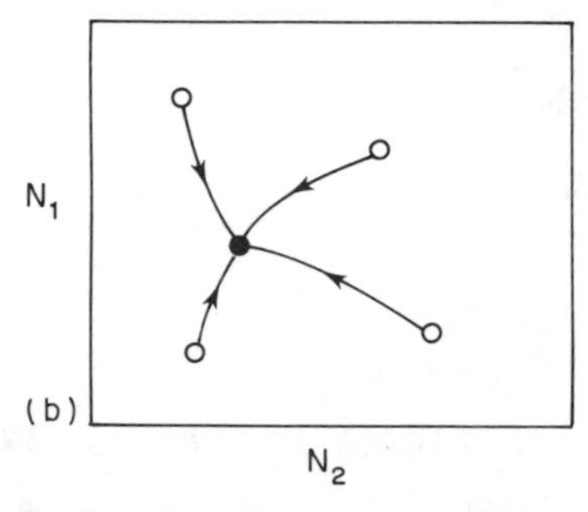

Figure 4.1 (a) Competitive exclusion and (b) stable coexistence illustrated in an N_1/N_2 phase plane. O—Starting point (arbitrary). ●—Finishing points

than two-dimensional) taken up by variables describing the state of the system. In this latter case, time appears 'internally' as a series of arrows (or trajectories or vectors). There is no special name for the first of these 'pictures'—they are just graphs of whatever variable is monitored against time. The second kind of picture is called a phase space or, in the special case of two dimensions, a phase plane. I will adopt the phase–space approach here, the 'graphical' one in the following section.

The concept of a phase space is a very general one, which can be used to describe the behaviour of any system, not just population–biological ones. It is the way in which cyberneticists picture systems (see e.g. Pask, 1961). In the context of interspecific competition, the axes of a phase–space are the densities (N_1, N_2, N_3, . . .) of the species concerned. I will concentrate as usual on the simplest situation and hence on the N_1/N_2 phase plane that is applicable to two-species systems. The necessary concepts can most easily be pictured in the two dimensions of a phase plane, and they are not fundamentally different in this special case to their general n-dimensional equivalents.

In the context, then, of an N_1/N_2 phase plane, an equilibrium point is any point—i.e. any joint set of densities for the two species—that will remain static over time. Equilibria may be stable or unstable; the former are characterized by the return of displaced values to them. If displaced values return from anywhere in the phase plane the equilibrium is globally stable. If, on the other hand, values displaced to some parts of the phase plane do not return to the equilibrium, but display some other sort of behaviour (such as a 'decay' to zero on one of the axes), then the equilibrium is described as locally stable. Much of competition theory is concerned with the conditions that switch a system from one without a non-trivial stable equilibrium (competitive exclusion: see Figure 4.1a) to a system that has an equilibrium that is globally stable ('global coexistence': see Figure 4.1b).

There are, of course, all sorts of complications to the contrast between the two fundamentally different systems illustrated in Figure 4.1, some of which are as follows:

1. Competitive equilibria may only be locally stable, as in the case of some of Gause's (1935) experiments (see next chapter).
2. It is possible to end up with a persistent cycle of values rather than a single equilibrium value. Such 'limit cycles' may, like equilibrium points, be stable or unstable. They have been invoked in both competitive (Armstrong and McGehee, 1980) and polymorphic (Clarke, 1976) systems.
3. Return to equilibrium following perturbation can occur in different ways. Systems where values change in discrete jumps have a tendency to overshoot equilibria—as have continuous systems with time-lags. In population ecology this is called overcompensation (see Varley *et al*., 1973).
4. If equilibria are only locally stable, it is possible to have a system which exhibits several different stable equilibria or stable and unstable zones in the same phase space.
5. Whatever are the deterministic stability characteristics of a system, they will be liable to stochastic effects where either laboratory or natural populations are concerned. In particular, equilibria with N_1 and/or N_2 close to zero will be prone to collapse as a result of stochastic processes.
6. A stable equilibrium may itself be 'unstable' in the sense that it may disappear if there are slight changes in the details of the system from which it arises (be it a mathematical model or a real biological system). The tendency for equilibria to persist despite such changes can be referred to as robustness or structural stability; some aspects of the robustness of competition models are discussed by May and MacArthur (1972), among others.

I will be mostly concerned, in the following chapters, with the basic contrast between unstable and stable systems, as illustrated in Figure 4.1, and will be less concerned with the precise nature of the equilibrium. The emphasis will be on biological mechanisms that cause systems to be stable. It is true that some mechanisms tend to cause certain kinds of stability (e.g. non-transitive competitive abilities lead to limit cycles—see Gilpin (1975)—while niche differentiation does not). However, few enough *experiments* on competition have revealed any kind of stability, and those that have have shown what appear to be equilibrium points whose stability is either global or at least extensive, i.e. local but covering a large proportion of the phase plane. Also, it should be admitted that the 'messiness' inherent in most real population data (including that deriving from laboratory populations) often makes it difficult to give a precise description of the stability characteristics of the system concerned, while it is usually not so severe as to prevent the making of a distinction between stability and a lack of it.

4.2 FREQUENCY DEPENDENCE

In theory, it would be possible to describe the behaviour of a two-allele polymorphism in exactly the same way that Figure 4.1 describes the behaviour of a system of two-species competition. The axes would now be the numbers of copies of alleles A_1 and A_2; alternatively, if there is complete dominance, a phase plane showing the joint numbers of the two phenotypes could be constructed. In practice, however, such a description is never used by population geneticists, and the reason for this is simple. Population numbers are rarely determined in studies of polymorphism; rather, the gene frequency or phenotype frequency is determined from a sample and presented in just that form, i.e. a *frequency*. In that case, a two-allele polymorphism becomes describable by a single relative-frequency variable rather than two density or number variables. Since it is impossible to have a one-dimensional phase space (or more correctly, impossible to draw a system's behaviour on this basis), the kind of representation suggested above cannot be employed, except where the gene-frequencies of two polymorphisms are being looked at in relation to each other. For visual presentation, gene-frequency data needs to be plotted against some other variable, and the obvious choice is time—though in another context gene-frequency along a spatial transect might be plotted. This brings us back to the graphical approach described earlier.

Figure 4.2 contrasts a transient (i.e. unstable) polymorphism with a stable or balanced one. (A neutral polymorphism would simply appear as a series of horizontal lines when plotted in this way.) It is clear that, in the stable situation, allele A_1 declines in frequency when it is common and increases in frequency when it is rare. This is merely a verbal description of the pattern of change shown in the figure. The question we now need to ask is: what causes this kind of pattern to arise?

The main (but not the only) two balancing mechanisms that act to produce stable polymorphism are heterozygous advantage and frequency dependence. Both give rise to the pattern shown in Figure 4.2(b), but they do so in different ways. I will now outline these two mechanisms, taking frequency dependence first.

Frequency dependence, or frequency-dependent selection, occurs when the

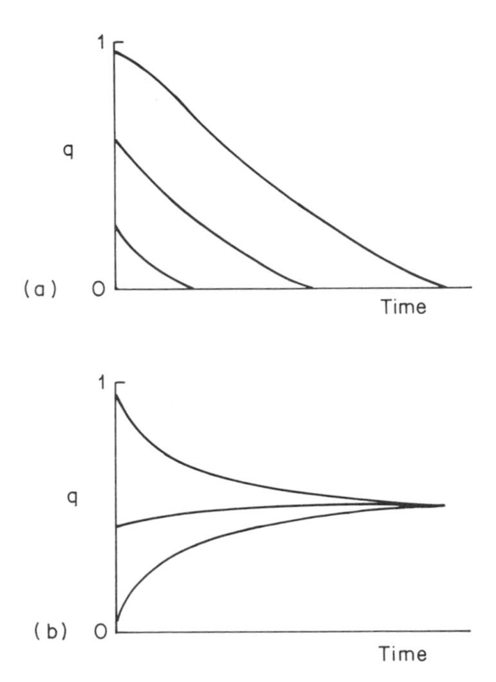

Figure 4.2 (a) Transient and (b) stable polymorphism (q represents the frequency of the A_1 allele)

fitness of a particular phenotype declines as its frequency increases. Various biological mechanisms produce such a state of affairs. These include niche differences between the phenotypes (Chapter 6), mating advantage of the rare phenotype (Petit, 1968) and apostatic selection (Clarke, 1962), i.e. predators concentrating disproportionately heavily on the commoner phenotype. [Apostatic selection is the genetic equivalent of predator switching (see Murdoch, 1969); that is, if alternative prey forms are phenotypes within a species the phenomenon is called apostatic selection, but if the prey are different species then it is referred to as switching.] The production of frequency dependence by these and other mechanisms has been reviewed by Murray (1972, Chapter 3) and Ayala and Campbell (1974). A general treatise on frequency dependent selection will appear shortly (B. Clarke, in preparation).

The production of the pattern shown in Figure 4.2(b) through frequency dependence can in a sense be direct or indirect. It is 'direct' when the organism concerned is haploid, in which case allele and phenotype coincide. In that case the entity whose frequency is seen to rise or fall in the figure is also the entity whose fitness is varying inversely with its frequency. However, in a diploid organism, allele and phenotype are clearly not the same thing, and in this context it is impossible to talk of the fitness of an allele, because it is phenotypes rather than alleles that survive or die and do or do not have offspring. Consequently the link between frequency dependence and the pattern shown in the figure is indirect. To put this another way, the graphical pattern *is* frequency dependence in the case of the haploid organism whereas it is a *result* of frequency dependence in the diploid case. Moreover, in diploids such a pattern may arise in the absence of frequency dependence through the action of a distinct mechanism, namely heterozygous advantage.

In contrast to the variable fitness values that characterize frequency dependence, heterozygous advantage involves *fixed* fitnesses with, as the term implies, that of the heterozygote being the highest of the three. Given such a situation, it can be shown that, in contrast to the directional selection described in Chapter 3, there is a stable polymorphism with equilibrium frequency of the A_1 allele

$$\hat{q}_1 = \left(\frac{s_{22}}{s_{11} + s_{22}}\right) \tag{4.1}$$

where s_{11} is the selective coefficient acting against the genotype A_1A_1 and s_{22} that acting against A_2A_2.

The way that heterozygous advantage produces the pattern shown in Figure 4.2(b) can be thought of as follows. Although the fitnesses of the *genotypes* are fixed, when either allele is rare most copies of it are found in heterozygous form and thus are carried by a fit genotype and so increase in number relative to the other allele, most of whose copies are carried in homozygous form. In this way, the pattern of *allele*-frequency change shown in the figure is produced without frequency-dependent fitnesses of genotypes or phenotypes.

One difference between these two balancing mechanisms that should be made abundantly clear is that heterozygous advantage (of any degree) *must* lead to a stable polymorphism, with an equilibrium gene frequency determined by equation 4.1; while frequency dependence *can* lead to stability but need not necessarily do so. To produce an equilibrium, the relative fitness of a variant must, in the course of its descent along an increasing frequency scale, cross the 'line of equal fitness' as shown in Figure 4.3(b). If it does not do so, as in Figure 4.3(a), then no equilibrium will ensue.

So far in our discussions of stability, we have dealt with stability of coexistence in terms of phase planes, that of polymorphism in terms of frequency/time graphs. I have stressed that it is usually impossible to plot polymorphic data on a phase plane. However, it is possible to plot species data as a frequency/time graph, and indeed this has often been done (e.g. Ayala, 1969, 1971). An advantage of doing so is that we see clearly the analogous nature of stability in the two types of situation. A disadvantage, which potentially applies to genetic, as well as ecological, data, is that the same relative frequency can be arrived at at different overall densities, and the system may behave differently in these different density conditions. If so, graphs of frequency against time may show at one stage a frequency increasing from a starting point of (say) 60 per cent and at another stage a frequency

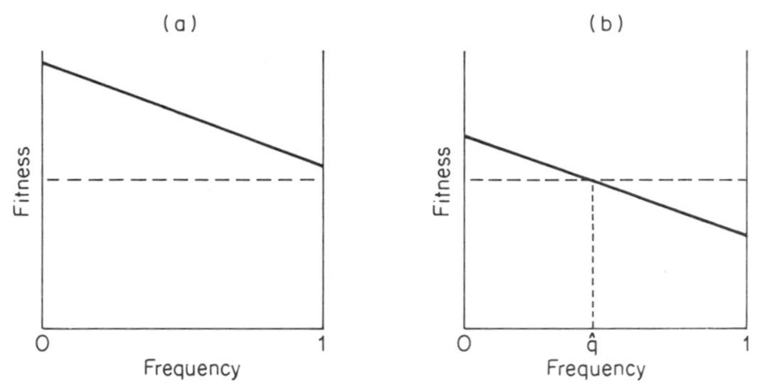

Figure 4.3 Two idealized patterns of frequency dependence: (a) No equilibrium. (b) Stable equilibrium at $\hat{q}$

decreasing from the same point, suggesting indeterminate system behaviour. If this happens, our graph is an insufficient description of the system concerned, as it contains too few variables to give us a true picture of what is happening. (See Lewontin, 1974 for a general discussion of this problem.) Because of this possibility, it would seem preferable to plot data on interspecific competition in the form of a phase plane. However, we must be clear that, whatever way the data are plotted, stable coexistence of competing species requires there to be frequency dependence, or, to state it more fully, frequency-dependent competitive abilities. There may be complicated patterns of frequency dependence if, as stated above, the same frequency achieved at different overall density levels gives rise to different system behaviour. Also, one must make the same restriction as in the polymorphic situation that frequency dependence does not necessarily lead to an equilibrium. Nevertheless, *if the competitive ability of a species does not in some way decline as its relative frequency in the two-species mix increases, there cannot be stable coexistence*. I stress this because some ecologists talking in terms of niche differentiation seem to feel either that there is no connection with frequency dependence or (worse) that niche differentiation is somehow an alternative to frequency dependence. The true relationship is, as I hope to show in the next section, that niche differentiation is one biological cause of the statistical relationship that we call frequency dependence—possibly the commonest cause.

4.3 NICHE DIFFERENTIATION

In Chapter 1, I equated the niche with MacArthur's resource utilization function (RUF), which seems appropriate in the context of exploitative competition. Given this equation, niche differentiation means partially overlapping RUFs as shown in Figure 1.4; this situation is also frequently referred to as resource partitioning. Both terms are problematical, though in different ways. Differentiation sometimes implies entities *becoming* different, as it does in cell and population differentiation. Thus niche differentiation, which does not imply this, could easily be confused with niche divergence (see Chapter 7) which does. Partitioning implies a total separation of two entities—as in the partitioning of Ireland into North and South, the partitioning of a large laboratory into two small ones, or the partitioning of a mathematical set into a subset and its complementary subset. Such a usage in niche theory would imply completely disjunct RUFs, which is quite a distinct phenomenon from that illustrated in Figure 1.4. However, 'resource partitioning' is so deeply embedded in the ecological literature that to attempt to remove the term would be futile. Given that, we should at least devise other terms for the alternative situations that are *not* resource partitioning. Thus we could have *resource segregation* for completely disjunct RUFs and *resource identity* for precisely superimposed RUFs (Arthur, 1986). These exhaust the possible relationships between two niches of the same size and shape. If size and shape are allowed to differ, all sorts of other possibilities arise, such as *resource inclusion* where a small niche is contained within a bigger one. Shape differences can lead to a form of partitioning where neither species has a unique section of resource axis but 'partitioning' still occurs because two curves of opposite skew rest on the same base. In what follows, any comments made about 'resource partitioning' are applicable to this situation as well as to the more 'conventional' one shown in Figure 1.4. Having distinguished resource partitioning (alias niche differentiation) from segregation and identity, I will say little more about these other two situations. Resource segregation implies lack of competition and so is of little interest here, despite its undoubted commonness in randomly-chosen pairs of species. Resource identity, on the other hand, is bound to be exceedingly rare, and one could indeed argue that precise identity will never be found, at least where the niches belong to competing species rather than genetic variants within a species.

Having made clear what is meant by niche differentiation/resource partitioning, I now want to examine the effect of this phenomenon on interspecific competition, and in particular its effect in generating stable coexistence. (The comparable genetic situation will be considered shortly.) The discussion will be centred on Figure 1.4, and I will assume that both niches shown there have BPC = 0, i.e. within each population, individuals are considered to be identical. Certain aspects of niche theory, particularly coevolution (see Chapter 7) require consideration of the BPC, but the present discussion does not, hence the temporary simplifying assumption that it is zero.

The basic principle underlying the stabilizing effect of niche differentiation is that increasing the numbers of one population has a greater inhibitory effect on the *per capita* growth rate of that population than on the growth of the alternative one because the maximal depletion of resource occurs in precisely that zone of the resource axis that the species concerned specializes upon. Thus niche differentiation generates frequency dependence, and thereby (sometimes) stable coexistence.

Models which quantify this effect using the sort of background provided by Figure 1.4 tend to be collectively referred to as species packing theory (see e.g. MacArthur, 1970; May and MacArthur, 1972; McMurtie, 1976). An alternative genre of model is that in which there is a series of discrete resource types rather than a continuous resource spectrum. Examples include the models of Stewart and Levin (1973) and Lawlor and Maynard Smith (1976). Here, precisely the same principle applies: specialization on different resource types leads to frequency dependence and hence to stable coexistence.

While it is reassuring that continuous- and discrete-resource models tell us roughly the same thing, neither adequately reflects the kind of niches found in nature. It is common for individuals in a natural population to consume a variety of different resource *types* (e.g. different taxa) and to exhibit a preference *within* each type which can be expressed as an RUF. Thus the best description of a niche might be a series of RUFs, each occupying a different axis. (Exactly what niche axes may be will be dealt with in Section 4.5.) A theory of such 'complex niches', with discrete and continuous components, remains to be developed. This is a potentially very exciting area of future population biology, and one which would help to bring theory closer to reality.

Turning to the production of stable polymorphisms through niche differences, the same basic principle applies—namely that an increase in the number of one variant decreases the resources available to that variant more than those available to others, thus producing frequency dependence, and hence, sometimes, stable polymorphism. Unlike the niche theory of interspecific competition, 'polymorphic niche theory' started off with a discrete-resource approach (Levene, 1953). However, approaches to multiple-niche polymorphism using RUFs have also been made, and again, as with species, the two approaches yield a qualitatively similar answer—i.e. niche differentiation can produce stability.

The next two chapters will be largely concerned with experimental evidence for the production of stable coexistence (Chapter 5) and stable polymorphism (Chapter 6) by niche differentiation. The following sections take a brief look at stabilizing mechanisms other than resource partitioning (Section 4.4); and the question of what the x-axis of an RUF may actually be in practice—including the question of whether it may be space or time (Section 4.5).

4.4 OTHER STABILIZING MECHANISMS

In this section I will concentrate on mechanisms other than niche differentiation that can cause stable coexistence of competing species. Descriptions of polymorphic stabilizing mechanisms are much more widely available, and I have, at any rate, already described what is often perceived to be the 'main' one (heterozygous advantage: see previous section). Thus I will have little more to say here about stability of polymorphism, except to point out that, although it may arise from temporal variation in fitness values, the conditions required for this to happen are very restrictive (Haldane and Jayakar, 1963).

Strangely, the debate about the conditions required for stable coexistence of species has revolved not around stabilizing mechanisms but rather around the degree of ecological similarity of the species concerned, or, sometimes, on the applicability of a particular competition model, often the rather unrealistic Lotka–Volterra one (see Chapter 2). This state of affairs is unfortunate, and the aim of the present section (and Chapter 5) is to rectify it, by considering mechanisms first. The reasons for doing this are simple: first, the whole idea of comparing dietary similarity is only meaningful in cases where resource partitioning is the suspected stabilizing mechanism; second, it is only possible to model a competitive system in detail when a particular stabilizing mechanism has first been isolated for consideration. Indeed, one of the main problems of the Lotka–Volterra model, as noted in Chapter 2, is that interpretation of the conditions for coexistence ($\alpha < K_1/K_2$; $\beta < K_2/K_1$) is problematic because these conditions could result from niche differences or something entirely different—such as waste products that are more harmful intra- than inter-specifically.

In addition to niche differentiation/resource partitioning, the following mechanisms are capable of causing stable coexistence of species undergoing exploitative competition:

1. *Spatial aggregation.* If only a single resource type exists but this is broken up into a number of discrete patches, various population processes, such as aggregation of the superior competitor, will produce stable coexistence. This situation has been examined by several authors, including Shorrocks *et al.* (1979) and Atkinson and Shorrocks (1981). However, it is difficult to draw a clear distinction between this proposed stabilizing mechanism and resource partitioning, as will be discussed in Section 4.5.
2. *Resource presentation/depletion mechanisms.* If a single resource type is present in a single spatial patch, stable coexistence can still result under certain conditions of resource presentation/depletion coupled with non-linear responses of the competitors to variation in resource densities (Stewart and Levin, 1973; de Jong, 1976; Koch, 1974a, 1974b; Armstrong and McGehee, 1976a, 1976b, 1980). However, such systems tend to exhibit stable *cycles* rather than a stable equilibrium *point*.
3. *Genetic feedback.* This mechanism involves coevolutionary change of the two competing populations. Basically, the idea is that each evolves an increased competitive ability when rare. Genetic feedback was proposed by Pimentel *et al.* (1965) on the basis of a particular set of experimental results. This proposal has been criticized by Levin (1971) and Arthur (1982a) but the mechanism, while unlikely, is at least conceivable. It should be noted that genetic feedback differs from other stabilizing mechanisms in that it involves competitive ability being negatively related to species frequency by an *evolutionary*, rather than an ecological, mechanism.

4. *Non-transitive competitive abilities.* When three or more species compete it is possible that competitive dominance networks of a non-transitive nature arise (e.g. A > B, B > C, C > A). Gilpin (1975) has shown that this phenomenon can cause coexistence, but as in the case of resource presentation/depletion mechanisms, the equilibrium takes the form of a stable cycle rather than a stable point. Also, it has been argued that in *exploitative* competition, competitive abilities are more likely to be transitive.

A few additional points need to be made about this list of stabilizing mechanisms. First, although they are phrased in terms of competing *species*, most of the mechanisms outlined could also operate within a species to produce stable polymorphism. This is true at least of the mechanisms listed as 1, 2 and 4 above. There is, of course, no equivalent of genetic feedback in the case of a single-locus polymorphism since an allele cannot evolve, but it could be argued that in a more complex kind of polymorphism, such as one involving a pair of chromosomal inversions, some sort of equivalent of genetic feedback is indeed possible. Second, as noted at the outset, the list of mechanisms given refers specifically to exploitative competition, i.e. situations in which there is a (−, −) interaction that is achieved through use of a jointly limiting resource. In cases in which one or both of these criteria are abandoned, a different range of mechanisms will apply. These cases include interference competition, apparent competition (Holt, 1977), amensalism (see Lawton and Hassell, 1981) and contramensalism (Arthur, 1986).

4.5 NICHE DIMENSIONS

I now consider the question of what exactly may constitute the x-axis of an RUF, i.e. what kind of variable may take the place of the rather abstract idea of a resource axis/spectrum/dimension in any particular example. I will continue to assume, throughout this section, that we need only consider a single such variable. The possibility that many dimensions should be considered simultaneously will be dealt with in Section 4.6.

In order to establish a reference point, let us start by considering the often-discussed case of a spectrum of seed sizes consumed by some species of bird. Here the resource axis takes the form of a physical variable (size) describing the *nature* of the resource. If two bird species have partially overlapping niches in relation to 'size of seed consumed', the situation is clearly describable as resource partitioning. If we substitute some chemical variable, also describing the nature of the resource, for size, then partially overlapping RUFs again constitute resource partitioning. In both these examples, there may of course be discrete or continuous variation (or both) but, as already noted, the mainstream of species packing theory has dealt with the continuous case.

The discussion so far is summarized in box (a) of Figure 4.4. Box (b) includes

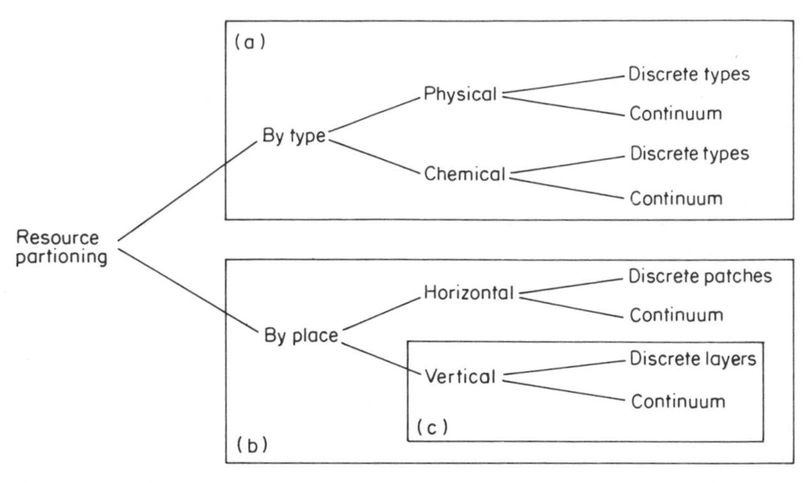

Figure 4.4 Types of resource partitioning. The *Paramecium* and *Drosophila* systems described in Chapter 5 both fall within box (c) (vertical partitioning). From Arthur (1986)

all situations where we have partitioning in relation to the *location* rather than the nature of the resource. Gause's (1935) experiments on *Paramecium*, which will be described in the following chapter, are widely interpreted as exemplifying vertical resource partitioning. As with partitioning by type, the species may be confronted with a discrete or continuous array of resource along the chosen dimension. In Gause's case, and in the *Drosophila* system described in Section 5.4, there was a continuous vertical distribution. A series of discrete vertical layers could also be envisaged, and here again a form of vertical resource partitioning could ensue.

Despite the inclusion of differential use of a vertical array of a single resource type under the heading of resource partitioning, differential use of a horizontal array has often been *contrasted* with resource partitioning, especially when the array is discrete, i.e. when there is a series of resource patches (see Shorrocks *et al*., 1979). Whether or not it is sensible to make this contrast depends on why the species make differential use of an array of resource patches. If, despite internal similarity, the patches differ for some 'extrinsic' reason—e.g. some are in exposed locations, others in shaded ones—then differential use of them by a pair of species may reflect repeatable behavioural differences between the species concerned, just as occurs in vertical partitioning in *Paramecium* or *Drosophila*. In this case, presented with several replicate ecosystems containing the same array of patches, those patches concentrated upon by species A (say) will be the same from one replicate system to another. In this case, there seems no reason to consider the situation to be something distinct from resource partitioning.

An alternative to this 'repeatable' situation is one where patches are identical with respect to intrinsic and extrinsic factors, and differential distributions of two species over an array of patches is essentially random, and caused simply by which individuals happen to find and utilize which patches (supplemented, in some cases, by certain kinds of aggregative egg-laying behaviour). Given several replicates of this kind of system, the patches in which species A is found will not be the same from one replicate to the next. Because of this random element, a case could be made for considering such situations to be distinct from 'true' resource partitioning. However, since different patches of what we perceive to be 'the same resource' will never be identical in all respects, it is doubtful whether differential aggregation of species over patches will ever be completely random and without any deterministic or repeatable component.

Having discussed all the possible types of resource partitioning given in Figure 4.4, I would like to briefly examine two further kinds of partitioning that are sometimes alluded to (see e.g. Schoener, 1974). The first of these is partitioning by habitat. I think it is best to avoid the use of this term, for the following reason. Different RUFs on a micro-spatial scale such as Gause's *Paramecium* may be described as partitioning by microhabitat (or, according to Pontin, 1982, as stratification). The microhabitats concerned are part of some larger overall area/volume being studied, which can be called the habitat—this may be anything from a forest to a test tube. Resource partitioning by habitat is only a meaningful phrase, in this context, when we are simultaneously considering two or more whole systems. At the current state of our knowledge of competition, there seems no reason to take such a complex approach.

The final kind of partitioning that deserves mention is 'temporal partitioning' where partially overlapping RUFs are based on a time axis. This is fundamentally different from all the other situations described above, because different points on the axis are not independent of each other. What I mean by this is that resource consumed at time t will not be available at time $t + \Delta$ (providing Δ is small). In contrast, the removal of a seed of diameter d has no depleting effect on seeds of size $d + \Delta$. Thus 'partitioning by time', rather than promoting coexistence, may promote competitive exclusion of the later-feeding species. (This sort of asymmetry parallels that occurring in certain situations where species differ in body size: see Wilson, 1975). What becomes important, with temporal partitioning, is the regeneration time of the resource. If this is sufficiently rapid, then temporal partitioning may indeed promote coexistence. This rather complex issue is dealt with by Haigh and Maynard Smith (1972); Schoener (1974) suggests that 'temporal partitioning' is relatively unimportant in natural communities.

So far, this section has been devoted to a discussion of the horizontal axis of the niche; the vertical axis has been assumed to be 'utilization', that is, some measure of the rate, or probability, of utilization at any point on the resource spectrum. Two brief points need to be made about such 'utilization'. First, it is not equivalent to fitness, and the labelling of the vertical axis of an RUF with fitness should be avoided. Second, it does not include any information on efficiency of conversion (into body tissue and/or offspring) of resources that are indeed utilized. This is a significant omission, and one which is rectified in some models of competition, including one discussed in Section 5.1.

4.6 *n* SPECIES AND *m* DIMENSIONS

Because this book is largely concerned with niche theory, and because this theory is not fundamentally altered when more than two species compete, I have largely confined my approach, at least in Parts I and II, to the simplest situations—i.e. where there are indeed only two competing species. This should not be taken to imply that systems involving three or more competitors are necessarily a straightforward extension of those involving just two. Indeed, one alternative stabilizing mechanism to niche differentiation—non-transitive competitive abilities—requires at least three species if it is to operate at all (see Gilpin, 1975, for the relevant theory, and Jackson and Buss, 1975; Buss and Jackson, 1979; and Jackson, 1979, for an apparent example of this mechanism). Whether the transition from two species to *n* species is a smooth one or not thus depends on the exact nature of the system under study. It seems unlikely that the niche-based coexistence described in the following chapter would be fundamentally altered if three or more species were involved, except that 'middle competitors' are overlapped on both sides by

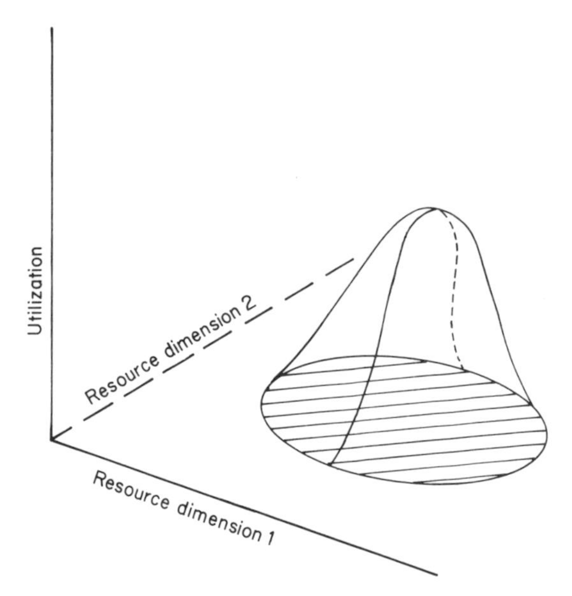

Figure 4.5 A resource-utilization 'dome' (a bivariate normal distribution). The base of the dome is shown as a shaded area on the 'resource plane'

the RUFs of competing species, and are thus in a sense more vulnerable than 'end competitors' (see MacArthur, 1972).

While niche theorists can be sceptical about the need to consider many species, they can hardly react in a similar way to the need to consider more than one niche dimension—i.e. more than one resource spectrum. It is true that the simple experimental systems described in the following chapter are characterized by separation in a single dimension and it seems likely that some natural systems are also essentially unidimensional. However, multidimensional separation of RUFs is likely to be the rule rather than the exception in nature. Whether or not this affects the argument being advanced in any particular case depends on whether it is being argued that an observed amount of niche differentiation in one (or a few) dimensions does or does not allow coexistence. If separation of RUFs along a single monitored dimension does allow coexistence, then no pattern of overlap/separation along unmonitored dimensions can preclude coexistence. However, if separation of RUFs on the monitored dimension does not allow coexistence, it may be that unseen multidimensional separation alters the situation completely.

One entirely new question which arises when RUFs with two or more resource dimensions are considered is that of whether utilization along one dimension is independent of that occurring along the other. Two dimensions with independent normal distributions of resource use give rise to a resource utilization 'dome' as shown in Figure 4.5. If utilizations along these two axes are not independent, then a variety of other shapes is possible (see Pianka, 1983, Chapter 7).

Chapter 5

Stable Coexistence and the Niche

5.1 BASIC THEORY

In Chapter 2 we examined the Lotka–Volterra (L–V) model of interspecific competition, and noted the conditions under which stable coexistence, rather than competitive exclusion, was predicted. I commented at that stage that, while these conditions were very simple mathematical entities (two inequalities: $\alpha < K_1/K_2$ and $\beta < K_2/K_1$), their biological interpretation was problematic because the model did not specify the *mechanism* of competition.

Subsequent theorists have attempted to get around this problem in one of two ways. Some have retained the L–V model but explicitly related the competition coefficients to overlap in RUFs, thus essentially restricting the model to situations of exploitative competition (e.g. May and MacArthur, 1972). Others have discarded the L–V approach completely and have built an alternative kind of model which contains parameters relating specifically to resource use *instead* of competition coefficients (e.g. Stewart and Levin, 1973; Lawlor and Maynard Smith, 1976). I will deal with these two approaches in turn.

Equating competition coefficients with RUF overlap, without any other modification of the L–V model, has the straightforward result that any degree of separation of RUFs short of complete overlap will permit stable coexistence. This in itself is not particularly informative. However, May and MacArthur (1972) also explored the effect of stochastic variation in population size, which took the form

$$K_t = \bar{K} + \gamma_t \qquad (5.1)$$

where γ was a 'Gaussian white noise parameter'—that is, an essentially random variation in K, with its values in successive time intervals being unrelated.

The prediction of this stochastic model is that the minimum value of d/w compatible with stable coexistence (the 'limiting similarity') increases as the amount of stochastic variation increases. However, the increase in d/w is far from linear: rather, it is rapid to begin with, plateaus for moderate amounts of variation, and then becomes rapid again as the environment becomes highly variable.

This relationship between limiting d/w and environmental variation has been looked at in two ways. First, some authors have drawn attention to the overall increase, rather than the particular *pattern* of increase in limiting similarity, and have pointed out that this may be important in explaining variation in species diversity among natural communities, such as the well-known latitudinal diversity gradient (see Pianka, 1966). Taking this line, Maynard Smith (1974) commented that 'the importance of this model is that it does give a first clue to understanding species diversity'. Other authors have focused on the 'plateau' of limiting similarity at a value of $d/w = 1$ for moderate amounts of stochastic variation. Since experimental populations in the laboratory experience environments that are neither purely deterministic (as in an algebraic model) nor extremely variable (as in many natural situations), the prediction of a limiting similarity of 1 should apply to them. If d and w can be measured, this opens up a way to test the model's prediction: stably coexisting populations with $d/w < 1$ would invalidate it, while coexisting-ing populations with $d/w > 1$ would be compatible with the proposed generalization. One such experimental test (Arthur and Middlecote, 1984a) is discussed in Section 5.4. Some further theoretical work on this issue has been done by May (1974a).

Turning to the alternative approach, namely models in which amounts of resource utilization are incorporated into the equations explicitly, rather than being 'hidden' in a generalized competition coefficient, we find that most models of this type have been based on discrete resources rather than on a continuous resource spectrum. Further, they have tended to concentrate on the simplest discrete-resource environment that will permit a degree of niche differentiation, namely an environment containing only two resources.

Two models of this kind are those of Stewart and Levin (1973) and Lawlor and Maynard Smith (1976), the latter stemming from MacArthur (1972). Although these were formulated for very different purposes (to model bacterial populations in a chemostat and to provide a basis for a generalized model of coevolution respectively), and although the symbolism used differs widely, the two models have much in common. The version given below is based on Lawlor and Maynard Smith (1976), though some of the symbolism has been altered to make it more readily comparable with the L–V model discussed in Chapter 2.

The two competing consumer populations are described by the following differential equations:

$$\frac{dN_1}{dt} = N_1(p_{11}e_{11}R_1) + N_1(p_{12}e_{12}R_2) - mN_1 \tag{5.2}$$

$$\frac{dN_2}{dt} = N_2(p_{21}e_{21}R_1) + N_2(p_{22}e_{22}R_2) - mN_2 \tag{5.3}$$

The first two components on the RHS of the equation represent natality due to feeding on the two resources (R_1 and R_2). The parameters p_{ij} and e_{ij} are, respectively, the probability of an individual of species *i* consuming an item of resource *j* per unit time, and the efficiency of conversion of that unit into new individuals. The third component in each equation represents mortality, which is given in its simplest (density-independent) form.

Neither consumer population appears explicitly in the equation for the alternative species, unlike the situation in the L–V model. However, interspecific interactions occur via the effects of the consumers on the resource populations, which are themselves modelled as follows:

$$\frac{dR_1}{dt} = R_1\phi(R_1) - (p_{11}N_1R_1) - (p_{21}N_2R_1) \tag{5.4}$$

$$\frac{dR_2}{dt} = R_2\phi(R_2) - (p_{12}N_1R_2) - (p_{22}N_2R_2) \tag{5.5}$$

Here, the last two components in each equation represent depletion of the resources by the consumers. The first component represents resource regeneration. The parameter ϕ can be any function—both MacArthur (1972) and Lawlor and Maynard Smith (1976) designate it as the logistic expression $r[(K - N)/K]$.

The conditions for stable coexistence in this and related models are far from simple. Lawlor and Maynard Smith do not consider these conditions—rather having stated the basic model they go on to use it to investigate coevolutionary change. MacArthur (1972) points out that the critical parameters are the p_{ij}s (or a_{ij}s in his version). The more different the two consumers are in their p_{ij}s, that is, the more one consumer concentrates on one resource and the other consumer on the alternative resource, the more probable is stable coexistence. Stewart and Levin (1973), using their related chemostat model, show that opposing specializations may cause coexistence but do not necessarily do so. This conclusion is based on a simulation approach.

A useful parameter in this context is the resource utilization ratio or RUR for species *i*, defined as $p_{i1}/(p_{i1} + p_{i2})$ (see Arthur, 1977, Chapter 8). This is a kind of discrete-resource equivalent of the RUF. An RUR has a scale from 0 to 1, with 0.5 indicating equal usage of both resources, and values on either side of this indicating some degree of specialization (on resource 1 if $0.5 < \text{RUR} \leq 1.0$ or on resource 2 if $0 \leq \text{RUR} < 0.5$). I have simulated competition according to the MacArthur–Lawlor–Maynard Smith model in the simplest situation—that in which the consumers differ only in p_{ij}s. In this case it turns out that as long as RUR_1 is on one side of 0.5 and RUR_2 on the other, stable coexistence will ensue. That is, any degree of opposing specialization is sufficient to promote coexistence. This is comparable to the prediction of the L–V model, interpreting the αs and βs in terms of niche overlap.

There are several difficulties faced by the experimentalist intent on testing the models described above. First, the conditions for coexistence are not known in algebraic terms for some of them, and conditions established from simulations based on a restricted set of numerical values are a poor substitute. Second, despite being more complex than the basic L–V model, all the models described in this section are too simple to be accurate models of most experimental systems. All, for example, exclude age and sex differences. In fact, they are really just more elaborate strategic models which help us to think quantitatively but should not be expected to be realistic abstractions of the experiments described later in this chapter.

An alternative approach for the experimenter is to try to determine which of the stabilizing mechanisms described in Chapter 4 is operating in a given system. Indeed, I would argue that a strategy of attempting to identify the correct mechanism is more sensible than attempting to determine the correct model. After all, the former ultimately tells us something biologically meaningful, while the latter does not. I will proceed, in subsequent sections of this chapter, to take a mechanism-based approach. As we will see, this sometimes leads to the testing of a model *after* one particular mechanism has been shown to be acting; but I regard this as an optional extra in a programme of competition experiments, not the ultimate goal.

5.2 THE IDEAL EXPERIMENT

If an experimental programme is to be devised to discriminate between the alternative biological mechanisms that can cause stable coexistence, then the form that programme should take depends on which stabilizing mechanism it is sought to demonstrate. The 'ideal experiment' described below is only ideal in the context of attempts to show that a particular state of stable coexistence is caused by niche differentiation/resource partitioning. Consideration of the ideal experimental design in such a study helps to reveal that many previous experiments that have been widely interpreted as 'proving' the efficacy of resource partitioning as a stabilizing mechanism (e.g. Gause, 1935; Crombie, 1945) are seriously lacking and cannot be regarded as conclusive.

The primary requirement, of course, is for a state of stable coexistence to be revealed—in the absence of this, we cannot proceed at all. Given this as a starting point, we are then faced with the task of trying to demonstrate that the coexistence is caused by differential use of a heterogeneous resource. If the coexistence is so caused, it should be possible to eliminate it by reducing the resource heterogeneity to a level at which differential use of it by the competing species is impossible. (There is no point in talking of a homogeneous resource—in reality there is no such thing.) Thus we need to demonstrate competitive exclusion in a second system that is identical to the first in all features except the level of resource heterogeneity.

Of course, in practice it is rarely or never possible to alter a single factor in an experimental design, leaving all others unchanged, despite the widespread notion of a 'control' suggesting that this can readily be done. It is easy to illustrate this assertion in the context of an attempt to alter resource heterogeneity but nothing else. Suppose a laboratory population is maintained in some sort of cage which contains two pots of resource—one of resource A and one of resource B. This two-resource system can be changed into a one-resource system by removal of the pot of B. However, while resource heterogeneity has been halved by this alteration, so has the *total amount of resource* available. Of course, we could replace the missing pot of B with a second pot of A, but now the variable 'quantity of resource A' has doubled in value. Essentially, changes in heterogeneity are always associated with changes in other variables.

Unless we refine our experimental design somewhat, we may fall into the trap that while we attribute the collapse of the state of coexistence on reducing the resource heterogeneity to the prevention of the resource partitioning mechanism, the collapse has actually been brought about by the prevention of another unsuspected mechanism through the alteration of a variable associated with the level of heterogeneity. While it may never be possible to eliminate this sort of interpretation completely, the case for an explanation in terms of resource partitioning can be greatly strengthened by quantification of both the proposed form of partitioning and the levels of resource heterogeneity, and by relating these quantities to each other. Again, a simple example will help to illustrate what I mean. Consider two RUFs of identical size and shape, each having a range equal to three units on the resource spectrum. Suppose that our 'heterogeneous' environment allowing coexistence has five units available and the two RUFs overlap by a single unit. If the amount of heterogeneity is reduced to three units of resource spectrum, the RUFs, if they remain similar in form, must overlap completely; thus resource partitioning is precluded. Given competitive exclusion in this second system and stable coexistence in the first, the demonstration that resource partitioning occurs in the system producing coexistence but not in the other makes the case for partitioning being the stabilizing mechanism very strong. I would argue that the onus then falls on the sceptic to outline some concrete alternative explanation of the overall set of results if his scepticism is to be taken seriously.

In summary, then, for conclusive demonstration that resource partitioning is the cause of an observed state of stable coexistence, the following demonstrations are necessary:

1. Demonstration of stable coexistence in a system with a given level of resource heterogeneity.

2. Demonstration of competitive exclusion in a less heterogeneous system.
3. Quantification of RUFs and demonstration that significant separation is possible in (1) but not in (2).

Even this fairly rigorous protocol is not exhaustive. There are some fundamental demonstrations that should underlie it—such as demonstration, through comparison with monocultures, that the species are indeed competing. Also, strictly, the results of system (2) should be analysed to ensure that competitive abilities were not frequency dependent—though a problem here is that if the competitive exclusion is rapid, the power of tests for frequency dependence is low. If they were frequency dependent, then it could be argued that the difference between (1) and (2) is merely that between Figures 4.3(a) and 4.3(b), and that we are not dealing with the stabilizing mechanism at all, but only with some peripheral factor influencing whether that mechanism does actually produce an equilibrium.

Having identified the criteria for demonstrating the production of stable coexistence by resource partitioning, we now turn to examine some case studies of particular systems to see the extent to which the criteria were satisfied. The following section deals with the work of Gause (1935). Section 5.4 deals with two series of experiments on *Drosophila*—one conducted by Ayala (1969, 1970, 1971), the other by myself and Judith Middlecote (Arthur and Middlecote, 1984a).

5.3 GAUSE'S *PARAMECIUM*

In Chapter 2, we examined the results of Gause's (1934) experiments with the species *P. aurelia* and *P. caudatum*, all of which led to (or at least towards) competitive exclusion of the latter species. We now turn to Gause's (1935) later experiments in which both of the above two species (separately) competed with *P. bursaria* (see also Gause, 1936, 1937; Gause and Witt, 1935).

The outcome of both of these mixed cultures was—in contrast both to Gause's earlier experiments and to the 'typical' laboratory experiment on interspecific competition—stable coexistence of the two species concerned. These coexistences are illustrated by the phase plane diagrams of Figure 5.1; these have been re-drawn from Gause (1935) and it should be stressed that they are Gause's '*idéalization*' of the results, not the actual data (which are somewhat messier).

Figure 5.1 shows two important features of the systems studied. First, some combinations of starting numbers (top parts of diagrams) led to a single stable equilibrium. Second, this equilibrium was local rather than global (see Chapter 4), as it failed to attract the trajectories starting from other combinations of initial numbers of the two species (lower right parts of phase plane diagrams). Gause refers to the sections of the phase planes that are outside the region of attraction of the 'main' equilibrium as special zones ('*zones particulières*'). It is not entirely clear what happens in the special zones—Gause appears to indicate the existence of multiple alternative equilibria,

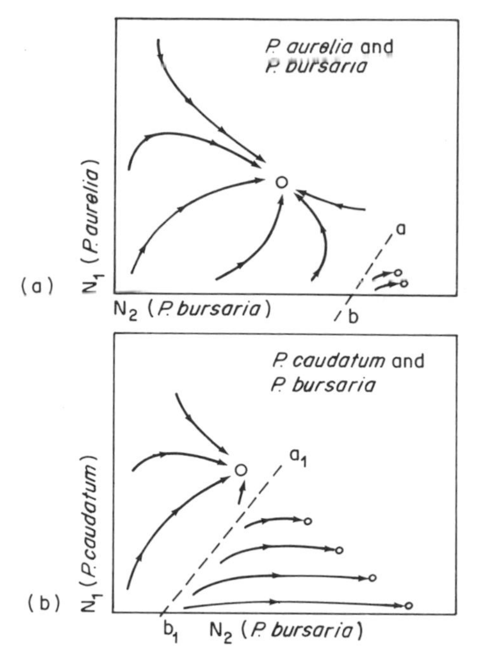

Figure 5.1 Phase plane diagrams showing the outcome of competition between (a) *Paramecium aurelia* and *P. bursaria*, and between (b) *P. bursaria* and *P. caudatum*. The 'special zones' are to the right of the dashed lines ab and a_1b_1. From Gause (1935)

which seems rather unlikely. Nor is it clear what causes the special zones, though Gause attempts an explanation based on the effects of waste products. However, one thing that is clear is that the special zone is much larger, in the case of competition between *P. bursaria* and *P. caudatum*, than it is in the *P. bursaria*/*P. aurelia* system.

Special zones aside, it is apparent that Gause has demonstrated stable coexistence and therefore has fulfilled the first requirement of the 'ideal experiment' described in the preceding section. What of the other two requirements? In fact, neither of these are met. With regard to the mechanism of coexistence, Gause claimed that it was niche differentiation with *P. bursaria* feeding at a deeper level in the culture tubes than either of the other two species. Unfortunately, though, Gause did not quantify the RUFs on a depth axis (or any other) and so we have no data to back up the assertion that there was indeed resource partitioning. This leaves open both the possibility that there was resource identity and the coexistence was achieved by some other (unknown) mechanism; and also the possibility that there was resource segregation, and so a lack of exploitative competition. It is clear from Gause's comparisons of equilibrium numbers in mixed cultures and monocultures that there was indeed a competitive interaction—but this could be of the interference type (via waste products)—a suggestion made by Case and Gilpin (1974). Gause's (1935) comments on the niches of the three species of *Paramecium* involved in his experiments seem to favour the latter explanation. For example, with respect to *P. bursaria* and *P. caudatum*, he says '*les niches écologiques des deux espèces ne coincident point*' which could be interpreted as implying resource segregation.

One additional point should be made about the proposed form of resource partitioning. Some of Gause's experiments involved a mixed food supply of bacteria and yeasts. In these experiments the yeasts tended to sediment towards the bottom of the tubes. Thus the proposed bottom feeder, *P. bursaria*, would also be a yeast specialist, and we would have resource partitioning along two (non-independent) axes. However, it is clear that specialization on bacteria/yeast is a secondary phenomenon, not necessary for stable coexistence, because some of Gause's experiments with *P. bursaria* and *P. caudatum*, and all of his experiments with the other species pair, involved yeast as the sole food source, and coexistence still occurred.

So far, we have seen that criterion (1) of our 'ideal experiment' was met in Gause's work, while (3) was not. This leaves (2), namely the demonstration of competitive exclusion in a system that is less heterogeneous than the one in which coexistence took place. Given the proposed stabilizing mechanism—resource partitioning by depth—the obvious kind of culture tube to set up for this demonstration is one with a very restricted depth of resource, which would not allow significant separation of the species in the vertical dimension. Regrettably, it appears that these experiments have not been done. [The fact that competitive exclusion occurs in a different species *pair* of *Paramecium* (see Chapter 2) is not in itself very informative.]

The inevitable conclusion, then, is that while Gause (1935) very definitely did demonstrate stable coexistence of competing species, itself an important demonstration at the time, he equally definitely did not show the mechanism by which that coexistence was brought about. He suggested a mechanism, and a plausible one at that, but the gulf between a plausible suggestion and a conclusive demonstration is a very wide one.

5.4 *DROSOPHILA* SYSTEMS

Ayala (1969, 1970, 1971) revealed three separate cases of coexistence involving three different pairs of *Drosophila* species. It is difficult to tell whether the coexistences described in the first two of these studies were stable, as only a single starting frequency (50 per cent of each species) was employed. I will thus say nothing further about them in the present context.

In his 1971 paper, Ayala showed a clear case of stable coexistence of *Drosophila pseudoobscura* and *D. willistoni*, in which three divergent starting frequencies converged to a common equilibrium value (see Figure 5.2). From

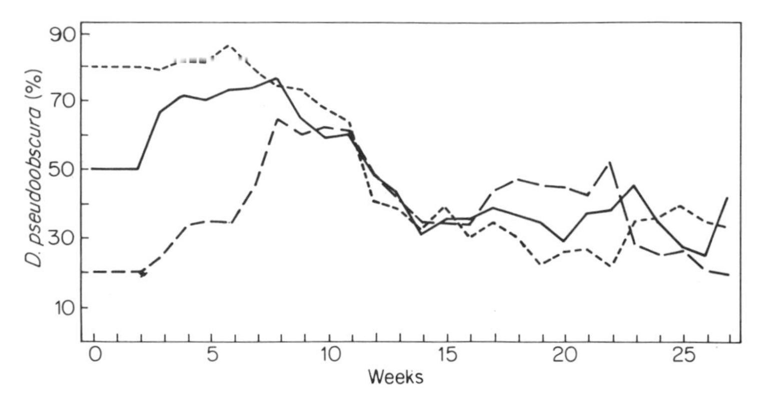

Figure 5.2 Stable coexistence of *Drosophila pseudoobscura* and *D. willistoni*. Reproduced with permission from Ayala (1971). Copyright 1971 by the AAAS

the data available we might predict that the equilibrium is globally stable; but of course there may be other combinations of starting numbers which would lead to a different result. There is a general message here: unlike theoreticians, experimentalists can never be certain that a given equilibrium is globally stable.

So far, Ayala's (1971) study resembles that of Gause (1935) in revealing a clear case of stable coexistence—and again, I will concentrate here on the central question of the *mechanism* of coexistence. Strangely, Ayala does not address this issue. He does, correctly, attribute the observed stable equilibrium to frequency-dependent competitive abilities. Indeed, the last sentence of his 1971 paper reads as follows: 'Frequency dependence leads, therefore, to a stable coexistence of the two competing species'. However, this evades the question of *what biological mechanism caused the frequency dependence*, and it is not clear why Ayala did not go on to consider this question. Whatever the reason, its lack of consideration means that the take-home message of Ayala's (1971) experiment is the same as that of Gause's (1935) experiments, namely that we know that the species concerned coexisted, but we do not know why.

An interest in picking up where the above studies left off and conclusively demonstrating the operation of a particular stabilizing mechanism led me to initiate a series of long-term competition experiments with pairs of *Drosophila* species. The first series of experiments (Arthur, 1980a,b) involved the sibling species *D. melanogaster* and *D. simulans*. Like most previous experiments on this species pair carried out by other workers, these experiments resulted in competitive exclusion, despite some being conducted in two-resource environments deliberately engineered to promote coexistence (Arthur, 1980b). They were thus completely uninformative about stabilizing mechanisms.

A second series of experiments involving *D. melanogaster* and *D. hydei* (Arthur and Middlecote, 1984a) were much more productive. Here, the

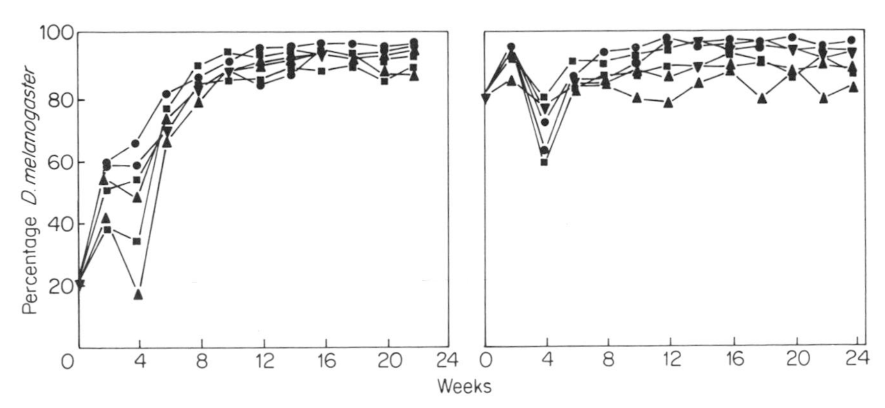

Figure 5.3 Stable coexistence of *Drosophila hydei* and *D. melanogaster*. Six replicate cages were started at each of two initial frequencies (20 per cent and 80 per cent *D. melanogaster*). Equilibrium frequency is at about 90 per cent *D. melanogaster*. Reproduced by permission of Academic Press Inc. (London) from Arthur and Middlecote (1984) *Biol. J. Linn. Soc.*, **23**, 167–176

species coexisted stably in multi-generation experiments as shown in Figure 5.3. Because the equilibrium frequency of *D. hydei* was rather low, giving a 'fringe' equilibrium which is difficult to distinguish from a slow trend towards competitive exclusion, a stock of *D. melanogaster* was inbred for several generations to weaken it and then used in a further competition experiment

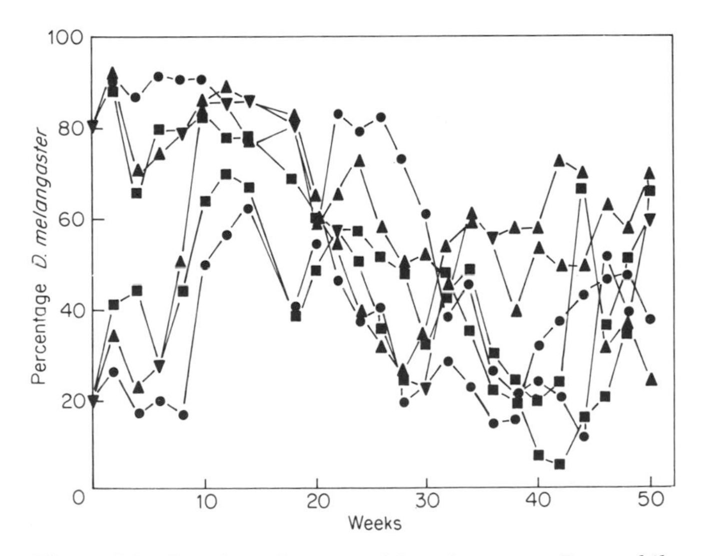

Figure 5.4 Results of competition between *Drosophila hydei* and *D. melanogaster* when an inbred strain of the latter species was used. Note the lower equilibrium frequency of *D. melanogaster* compared to Figure 5.3. Reproduced by permission of Academic Press Inc. (London) from Arthur and Middlecote (1984) *Biol. J. Linn. Soc.*, **23**, 167–176

with *D. hydei* (Figure 5.4). By the time this experiment had run for as long as the earlier experiments (22–24 weeks) the frequency of *D. melanogaster* had equilibrated at about 60 per cent (rather than about 90 per cent). On being continued up to 50 weeks, the experiment continued to show coexistence of the species, though the frequencies fluctuated quite considerably.

These experiments were carried out in population cages each with six resource bottles screwed into their undersides (for photographs of cages see Arthur, 1986). The above result, namely coexistence, occurred when each resource bottle was effectively full of the single resource-type used (instant *Drosophila* medium or IDM). The depth of resource in each bottle was 2–3 cm, and the amount of IDM that was hydrated to achieve this depth was 5 g. Inspection of resource bottles in competition cages and stock cages suggested that *D. hydei* larvae fed lower down in the resource bottles than *D. melanogaster* larvae. In other words, it seemed as if there was vertical resource partitioning, just as proposed by Gause (1935) in his *Paramecium* experiments.

In order to test the hypotheses (a) that resource partitioning was taking place, and (b) that it was responsible for the observed coexistence, two further

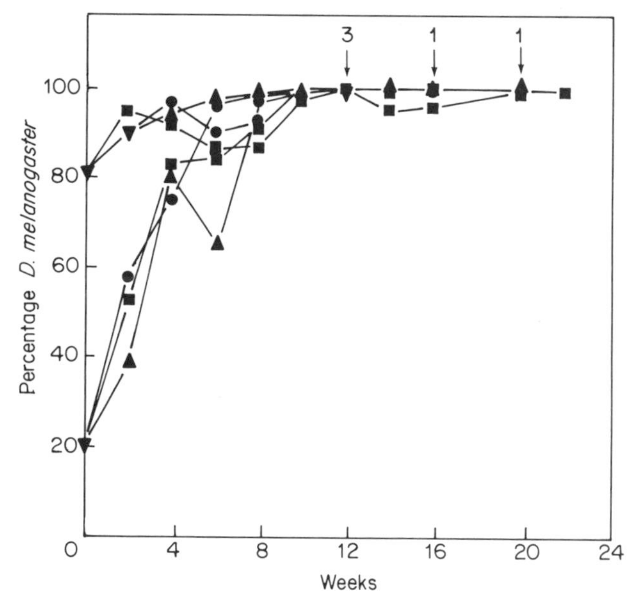

Figure 5.5 Competitive exclusion of *Drosophila hydei* by *D. melanogaster* in a shallow-resource environment. Arrows indicate extinctions of *D. hydei*; numbers above arrows indicate how many cages experienced extinctions at the times concerned. The sole cage remaining at the end of the experiment had more than 99 per cent *D. melanogaster*. Reproduced by permission of Academic Press Inc. (London) from Arthur and Middlecote (1984) *Biol. J. Linn. Soc.*, **23**, 167–176

kinds of experiment were carried out, as follows. First, the amount of IDM per bottle was reduced to 1.5 g, giving a depth of around 0.8 cm. This is about the same as the length of a large larva. Since larvae tend to feed in a vertical position, this very restricted depth of resource (which becomes even less as the food gets used up) effectively precludes partitioning by depth. A multi-generation experiment run under these conditions resulted in competitive exclusion (Figure 5.5). Moreover, competitive abilities were not significantly frequency dependent in this experiment, whereas frequency dependence was found in the experiments resulting in coexistence (Arthur and Middlecote, 1984a).

The other experiment that was set up was a short-term experiment conducted in vials with a 3 cm depth of resource, and designed to quantify the suspected interspecific difference in larval depth distributions. Distributions of larvae were monitored on the third and fourth days for *D. melanogaster* and on the fourth and fifth days for the slower-developing *D. hydei*. The results, summed over three replicate vials in each case, are given in Table 5.1. All interspecific comparisons are significant at $p < 0.001$ using a χ^2 2×2 contingency table.

While this series of experiments satisfies the three criteria identified in Section 5.2 for the 'ideal experiment', there is a significant omission. The original presentation of the results (Arthur and Middlecote, 1984a) contained no data on monocultures. Thus it could be argued that the species were not actually in competition. Comparisons of mixed cultures and monocultures over a single generation in vials did indicate a competitive interaction (Arthur, 1986, Section 6). However, extreme caution is needed in applying this conclusion to a multi-generation experiment conducted in cages.

The missing long-term monocultures have now been run, and comparison of these with the mixed cultures permits calculation of the interaction coefficients α (effect of *D. hydei* on *D. melanogaster*) and β (the converse effect). The results were:

$$\alpha = +2.41 \tag{5.6}$$

$$\beta = +0.79 \tag{5.7}$$

Table 5.1 Larval depth distributions for *Drosophila melanogaster* and *D. hydei*

Species	Day	Number of larvae	
		Lower section[a]	Upper section
D. melanogaster	3	4	141
	4	3	97
D. hydei	4	20	53
	5	75	32

[a] Two sections have equal depth—i.e. 1.5 cm each.

These coefficients are both positive, as is required of a (−, −) interaction! (See Chapter 2.) However, it must be admitted that there is no satisfactory way to test whether an individual competition coefficient is significantly different from zero. This matter has not been discussed by earlier workers (Gause, 1935; Ayala, 1969) who have calculated such coefficients and merely given their estimated numerical values. It is certainly not a matter which can be resolved here; but it is a topic which deserves further investigation.

Given that the experiments described above show a case of stable coexistence caused by resource partitioning, it is possible to proceed to examine May and MacArthur's (1972) model of limiting similarity. As noted in Section 5.1, this model predicts that we must have $d/w > 1$ for stable coexistence to be possible. Since the data in Table 5.1 were obtained by pooling larval numbers in 0.5 cm discs of medium, the mean and variance of larval depth can be estimated for each species, assuming that larvae were randomly distributed within each disc. From these data, d and w can easily be calculated and d/w ratios worked out. The results were 1.54 for the earlier set of data in the table and 2.59 for the later set. These values are clearly compatible with the May–MacArthur model's prediction, which means that they cannot be used to invalidate it; however, we can hardly claim, on this basis, that the model is adequate for competition between two multiple-life-stage species such as insects, which seems most unlikely.

5.5 FIELD STUDIES

The number of field studies conducted with a view to understanding the operation and importance of interspecific competition in natural communities is enormous. Yet despite this fact, few field studies provide us with clear evidence of stable coexistence and fewer still (arguably none) provide us with a definite conclusion as to the mechanism producing the stability.

This is a rather drastic statement to make, and it is worth clarifying it to reduce the risk of misinterpretation. I should stress that I am not attempting to write off all field studies on competition as a waste of time. Indeed many field investigations have a lot to tell us about other aspects of competition than the mechanisms underlying stable coexistence. Studies of species distributions in space and time provide circumstantial evidence—sometimes quite persuasive—of competitive exclusion, as discussed in Chapter 2. 'Explant' experiments in the field provide evidence of whether interspecific competition is taking place, and we will examine some of these in Chapter 8. Studies of variation in phenotypic characters in relation to allopatry/sympatry borders can, if rigorously undertaken, provide evidence of competitively-induced evolution (see Grant, 1972; Arthur, 1982a; and Chapter 7).

While these types of field study are informative in relation to specified aspects of the competitive process, they do not tell us anything about the causes of stable coexistence. What sort of study would be informative in this respect? First, we need to have evidence that there is stable coexistence. This would involve perturbation of population numbers (of both species) away from a putative equilibrium, and observation of their return to it over several generations. I am not aware of any such experiments having been carried out. In the absence of data from perturbation experiments, nothing can be said of the stability or otherwise of an observed coexistence. The mere fact that two species are found together in a particular habitat at a particular time is of no consequence. Coupled with positive results from a reciprocal explant experiment, such a coexistence could at least be described as a competitive one; but we would still be uncertain of whether it was stable or transient.

Given the difficulties of carrying out perturbation coupled with long-term monitoring of populations in the field, some studies have *assumed* stable competitive coexistence, and set out to quantify niche differences between the species concerned. When these are found (as they inevitably are), they are often interpreted as evidence of resource partitioning, which is then deemed to be responsible for the coexistence. There are many famous case-studies where an approach of this general kind was adopted, including 'MacArthur's warblers' (MacArthur, 1958), molluscs of the genus *Conus* (Kohn, 1959), and *Anolis* lizards (e.g. Schoener, 1968). While the niche differences revealed in these and countless other such studies *may* be the cause of coexistence, so many other interpretations are possible that it is imprudent to draw any firm conclusion from the data. Even more difficult to interpret are the studies in which niche differences are not measured but are inferred from morphological differences.

I will end this brief section with a plea that someone attempts a clear demonstration of stable coexistence of competing species in the field. With careful choice of organism, it should be within the bounds of possibility to get positive results. Then we would at least have a backcloth against which observed niche differences would be meaningful. It might even be possible to modify the habitat concerned to prevent the niche differences being manifested, which should lead to competitive exclusion. We would then have the 'ideal field experiment', which would in a sense supersede the laboratory studies described in the previous section.

5.6 THE PRINCIPLES OF COMPETITIVE EXCLUSION AND LIMITING SIMILARITY

The competitive exclusion principle is one of the few potentially general theories in ecology (Maynard Smith, 1972; Haigh and Maynard Smith, 1972), and the same could be said for the principle of limiting similarity that has arisen from it. Consequently these principles deserve careful attention. I will discuss their history, their wording (which can be crucial), the criticisms that have been levelled against them, the relationship between the two principles, and future experiments that could be conducted to examine their validity further.

Broadly speaking, the competitive exclusion principle states that two species with identical niches cannot coexist. When did a statement of this kind first arise in the literature on competition? It is always difficult to answer such a question with certainty, but it would seem that the first such statement can be attributed to Gause (1937), though the earlier theoretical work of Lotka (1925) and Volterra (1926) is also relevant, as is Darwin's (1859) proposal that competition increases in severity with increasing similarity of the competing forms. Certainly, the competitive exclusion principle (CEP) is frequently referred to as Gause's principle (or axiom or hypothesis or theorem). Some authors have objected to this because, they claim, Gause never actually formulated a general principle but merely provided us with examples of competitive exclusion from his early (1934) experiments on *Paramecium*. For example, Gilbert *et al*. (1952) state that Gause 'draws no general conclusions from his experiments, and moreover, makes no statement which resembles any wording of the hypothesis which has arisen bearing his name'. Nothing could be further from the truth. I have already given Gause's (1937) statement of the principle (in Section 2.2) but I will repeat it here for the reader's convenience: 'the steady state of a mixed population consisting of two species occupying an identical "ecological niche" will be the pure population of one of them, the one better adapted for the particular set of conditions'. While I do not advocate that we revert to calling the CEP Gause's principle, (largely because some earlier statement of it by another author may surface later), it should not be claimed that Gause did not give a clear and general statement of the principle. The confusion seems to have arisen because this statement was not contained in one of Gause's (1934, 1935) major publications. The critics mentioned above make reference to Gause's work, but do not refer to the 1937 paper in which the principle was stated.

In conclusion, then, we see foreshadowings of the CEP in the work of Darwin, Lotka, Volterra, and others, but the earliest known (at least to me!) statement of the principle itself was by Gause. Subsequent authors that have been instrumental in its further development include Lack (1947), Hardin (1960) Hutchinson (1965) and Armstrong and McGehee (1980), among many others. Further confusion has arisen because many authors have provided differently-worded versions of the principle. The differences are often slight, indeed often just a single word, but these can completely alter both what the principle is claiming and how (and whether) it can be tested experimentally. We will now examine, and comment upon, various formulations of the CEP.

First, it is quite clear that versions of the principle prohibiting the coexistence of species with *identical* niches are worthless because such species cannot be found. Even very simple laboratory environments containing a 'single resource' always permit significant differences in the feeding behaviour of the species—and sometimes these simple environments turn out to be astoundingly complex (see Arthur, 1986). In any natural environment, the idea of two species with identical niches is laughable.

While it is understandable that this problem has led some authors to replace 'identical' with 'similar', such a replacement causes other, and equally severe, problems. Since similar can mean whatever you want it to mean, the version of the CEP that prohibits the coexistence of species with similar niches cannot be experimentally tested. It is essentially a non-statement, which predicts nothing in particular and can evade any proposed invalidation (in the form of a demonstration of the coexistence of very similar species) because it can be claimed that the species concerned were different enough to be permitted to coexist under the 'principle'.

There are two ways out of this impasse. One is to specify some particular degree of similarity beyond which coexistence is supposed to be prohibited. This brings us to the principle of limiting similarity, which will be discussed shortly. An alternative is to formulate a version of the CEP which is not based on the degree of similarity at all, but rather is based on the mechanism of coexistence. I will now develop such a 'mechanistic CEP'. As will I hope be clear by the end of this section, the mechanistic CEP and the principle of limiting similarity make altogether different generalizations about coexistence.

Before stating the mechanistic CEP, I should stress that all versions of the CEP (and of limiting similarity) apply only to species in exploitative competition and that 'coexistence' implies stable coexistence. Now we have already seen (in Chapter 4) that there are at least five stabilizing mechanisms that could in theory produce a state of coexistence between species in exploitative competition, namely resource partitioning (alias niche differentiation), 'chance' spatial aggregation (see Section 4.5), genetic feedback, nontransitive competitive abilities and seasonal presentation of resource. The mechanistic CEP can be stated in a strong form, which asserts that:

> resource partitioning is the sole mechanism by which species undergoing exploitative competition reach a state of stable coexistence in natural or laboratory environments.

or in a weak form where 'predominant' replace 'sole'. The assertion then is that all the alternative kinds of stabilizing mechanism are figments of the imagination of theorists and never operate in reality (strong form) or at least that they are comparatively rare and unimportant (weak form).

Both forms of the mechanistic CEP are experimentally testable (the strong form more easily so), and neither can be rejected on the basis of any experiments conducted to date. What is needed to reject the strong form is an experiment that shows a state of coexistence caused by one of the stabilizing mechanisms other than resource partitioning. While some experiments have produced a state of coexistence in a system in which resource partitioning seems unlikely (Ayala, 1969; Levin, 1972), no such experiment has clearly shown what mechanism is promoting the stability, so resource partitioning cannot be ruled out.

We turn now to the principle of limiting similarity, which originated in the articles by MacArthur and Levins (1964, 1967), and, as Abrams (1983) comments in his review of the subject, is 'an outgrowth of the competitive exclusion principle'. Although the principle of limiting similarity tends not to be precisely defined, it basically asserts that there is some limit to the overlap in the RUFs of competing species that is consistent with their coexistence. Early ideas of some fairly general limit (e.g. $d/w = 1$; May and MacArthur, 1972) have given way to a suspicion that if there is a limit at all it is variable, and dependent on a variety of ecological factors (Abrams, 1983), which seems a more reasonable proposition.

It is important to note that the mechanistic CEP and the principle of limiting similarity have different frames of reference as well as different predictions. The domain of the mechanistic CEP is all cases of exploitative competition. Within this domain, it predicts that most (or all) cases of coexistence are caused by resource partitioning and not by some other mechanism. However, the limiting similarity principle (LSP) is not concerned with how coexistences split between those caused by resource partitioning and those caused in other ways. Rather, its domain is restricted to those that *are* caused by resource partitioning, and within this domain it attempts to predict how different the RUFs must be if resource partitioning is to 'work', i.e. to produce stability. Another way of putting this is that the LSP is irrelevant to cases of coexistence caused by mechanisms other than resource partitioning, because these mechanisms cannot be described in terms of RUFs.

Having reached this point, we can dismiss most criticisms of the CEP and LSP as misguided because they apply only to badly-worded or simplified versions of the principles. Complaints about early versions of the CEP (e.g. Cole, 1960; Pontin, 1982) are no longer valid in relation to the mechanistic CEP; also criticisms of a generally-applicable limit to the similarity of RUFs are no longer valid if we accept Abrams' (1983) proposal of a variable limit. Admittedly such a proposal is rather difficult to test; but the mechanistic CEP is testable, and I have already indicated the kind of experimental results that would invalidate the strong form of this principle.

I must admit to a suspicion that the mechanistic CEP is correct, though perhaps only in its weaker form. Whether it is important, however, is quite a separate issue. This really depends on whether exploitative competition is (a) common in nature and (b) important in the structuring of the communities in which it is found. These issues will be addressed in Part III. The final chapter of the present part of the book is devoted to the genetic equivalent of coexistence by resource partitioning, namely multiple niche polymorphism.

Chapter 6

Stable Polymorphism and the Niche

6.1 BASIC THEORY

The first model of polymorphism in a multiple-niche context was formulated by Levene (1953). Subsequent modifications of Levene's model have been provided by, among others, Maynard Smith (1966, 1970), Levins and MacArthur (1966), Strobeck (1974) and Yokoyama and Schaal (1985). The field of multiple niche polymorphism has been reviewed by Hedrick *et al.* (1976), Felsenstein (1976), Ennos (1983) and Hedrick (1986).

Levene's original model took the following form. The fitnesses of genotypes A_1A_1, A_1A_2 and A_2A_2 were permitted to vary between different niches, but in any one niche the fitnesses of the homozygotes were measured relative to that of the heterozygote. Thus we have the following situation:

Genotype	A_1A_1	A_1A_2	A_2A_2
Fitness	W_i	1	V_i

where the subscript i refers to the ith niche. The scenario Levene envisaged was one where mating was at random among the whole population, and the distribution of zygotes among niches was also random. There followed a period in which each member of the new generation was restricted to the niche in which it started (i.e. the environment was 'coarse-grained'); eventually, the adults emerged from their niches to mate. The situation would thus be exemplified by an insect whose eggs were distributed at random and whose larvae were capable of only very localized movement.

I have, in the previous paragraph, described the model in similar terms to those used by Levene (1953) and most of the later authors who have been active in this area. However, it should be stressed that 'niche' as used above and by Levene is not consistent with the usage of the term elsewhere in this book. In fact, the 'niche' with which Levene dealt was essentially a resource or microhabitat. Thus what we have is a heterogeneous environment with two or more resources. One important assumption of Levene's model is that there is separate density-dependent limitation of numbers in the different resource categories.

Levene designated the proportion of survivors in any generation that resulted from the ith resource as C_i (hence $\Sigma C_i = 1$). He was able to show that there would be a stable polymorphism if:

$$\sum \frac{C_i}{V_i} > 1 \quad \text{and} \quad \sum \frac{C_i}{W_i} > 1 \tag{6.1}$$

Taking the simplest situation of just two equally productive resources, so that $i = 1, 2$, and $C_1 = C_2 = 0.5$), we find that there will be a stable equilibrium if:

$$(0.5/V_1 + 0.5/V_2) > 1 \quad \text{and} \quad (0.5/W_1 + 0.5/W_2) > 1 \tag{6.2}$$

These conditions include both heterozygous advantage and (with different numerical values of the Vs and Ws) multiple niche polymorphism. For example, if V_1, V_2, W_1 and W_2 are all less than unity, then there is a heterozygous advantage in both resources. Alternatively, heterozygous advantage in one resource alone (say resource 1, i.e. $V_1 < 1$ and $W_1 < 1$) will promote coexistence without heterozygous advantage in the other resource, providing V_2 and W_2 are not too great.

For an interpretation of Levene's conditions solely in terms of specialization of genotypes on alternative resources, the numerical values need to be such that $V_1 > W_1$ and $V_2 < W_2$, but without there being heterozygous advantage in either resource. It can readily be seen that there are combinations of numerical values which satisfy these requirements (e.g. $V_1 = 1.5$; $V_2 = 0.5$; $W_1 = 0.5$; $W_2 = 1.5$).

Levene (1953) ends his paper by commenting that his model 'is obviously not realistic'. He notes, however, that making it more realistic is likely to make the conditions for polymorphism less rather than more restrictive. In particular, he draws attention to the fact that genotypes *choosing* to occupy resources in which they are fitter will make polymorphism more likely. This aspect of the model is developed by Maynard Smith (1966, 1970) who introduced a parameter measuring 'habitat selection'—a term actually used to mean choice of 'niche' (i.e. resource or microhabitat). Maynard Smith's (1966) measure of habitat selection was as follows for the simplified situation of two equally productive resources. A female lays a fraction $\frac{1}{2}(1 - H)$ of her eggs on resource 1, and $\frac{1}{2}(1 - H)$ on resource 2. If $H = 0$ there is no selection, while if $H = 1$ then habitat selection is complete. Maynard Smith found that systems with $H = 1$ had a range of fitness values giving stable polymorphism that was about twice as great as the range of such values when $H = 0$.

It is interesting to contrast the development of ideas on multiple niche polymorphism with those on stable coexistence of competing species through niche differences. In the ecological theory, the emphasis is very much on differences between species in their *choice* of resources. It is precisely this that MacArthur's RUFs describe in a quantitative way. There is little if any reference, in the ecological literature on this topic, to the relative fitnesses of the species in different resources (or at different points on the resource spectrum in the continuous case). It might be assumed that species would choose to utilize those resources which they are superior to their competitors at acquiring—and perhaps this assumption is often correct. (It is certainly not always correct—*Drosophila simulans* went extinct in competition with *D. melanogaster* in a two-resource environment because it failed to concentrate on the resource in which it was superior: Arthur, 1980b). This emphasis in the relevant ecological theory contrasts with the genetic counterpart, where Levene's basic model was phrased in terms of fitness differences, with actual *choice* of resource by a genotype being added as a secondary, modifying effect.

Since the Levene-type model is very general in its specification and does not invoke special characteristics found only in some particular kind of polymorphism, it is potentially applicable to *all* polymorphisms. In some ways this is a strength of the model, but looked at another way it is a weakness, in that no *mechanism* by which the niche differences could occur is suggested. Other authors, however, have proposed more restricted models of genetic polymorphism involving niche differences of a specific sort. An important model of this kind is that of Clarke and Allendorf (1979). These authors addressed the question of what kind of balancing selection might be responsible for maintaining the large amount of allozyme polymorphism found in natural populations. Since this is the biggest class of polymorphism whose means of persistence in populations is disputed, and since one of the main weaknesses of the 'selectionist' explanation of allozyme polymorphism is the lack of any kind of balancing selection that is generally applicable to enzyme loci, Clarke and Allendorf's (1979) model—which proposes just such a kind of balancing selection—deserves serious consideration.

Clarke and Allendorf argue that, provided certain fairly plausible assumptions are satisfied, it follows from the theory of enzyme kinetics that an allozyme with a higher value of V_{max} than its alternative form will also have a higher K_m. (V_{max} is the maximum reaction velocity when substrate is not limiting; K_m is the Michaelis constant, equal to the substrate concentration at which $V = \frac{1}{2}V_{max}$). The enzyme with the higher values will work faster when substrate is common, while its counterpart with lower values will be better when substrate is scarce. Such a system may operate to maintain polymorphism of an enzyme that acts upon an external substrate (found in one of the organism's resources) in the following way. If substrate is initially common, selection increases the frequency of the allele producing the enzyme with the higher values (E_H). As the frequency of E_H rises the substrate concentration is depleted more rapidly, causing, eventually, selection to reverse and favour E_L. This system is thus frequency dependent and can, under certain conditions, produce a stable equilibrium gene frequency.

There are several difficulties with Clarke and Allendorf's model, some of which are as follows. First, it assumes that the enzyme considered is rate-limiting in the metabolic pathway in which it is embedded. Second, it assumes that the rate of production of the ultimate product of that pathway is positively associated with fitness over all possible values. Even where these criteria are satisfied, the model will only work, in the form given above, for enzymes which digest external substrates. Finally, although the authors provided some data which lent support to the model, further data of this kind have not come to light.

I have spent some time describing the Clarke–Allendorf model despite the above difficulties because if the selectionist case for active maintenance of widespread allozyme polymorphism is to be viable, some *general* form of balancing selection is necessary. The model's importance stems from its proposal of such a form of selection.

Finally, I should point out that the Clarke–Allendorf model can only be considered a model of multiple niche polymorphism if one takes a rather broad view of what is meant by 'niche'. Basically, the enzyme genotypes concerned separate out on a 'resource abundance' axis rather than an axis of resource type. This means that, while the model has some features in common with a more conventional niche-differentiation system (notably the production of frequency dependence through ecological differences), it would perhaps be more accurately regarded as a genetic counterpart of the 'resource presentation/depletion' mechanism of stable coexistence (see Section 4.4) than as a genetic counterpart of niche differentiation in the strict sense.

6.2 LABORATORY EXPERIMENTS

Although the polymorphisms that are of most current interest are those involving allozymes, these are often difficult to work with because any selective differentials between allozyme genotypes at a particular locus are unlikely to be very great. This means that they are both difficult to detect and difficult to separate from selective differentials among genotypes at loci that are tightly linked to the one being studied. Thus, while any selective explanation of this class of polymorphism must ultimately be based on studies that overcome these difficulties, experimental illustration of the dynamics of polymorphism in general are often best made with polymorphisms whose constituent genotypes differ considerably in fitness. Some such polymorphisms are found in nature (the sickle cell polymorphism in man being an obvious example); others can be established by experimenters from combinations of genotypes which are not found in a polymorphic state in nature.

A study of a polymorphism of this latter sort is reported by Jones and Probert (1980), who established populations of *Drosophila simulans* polymorphic for wild-type and *white* alleles at the well-known X-linked locus affecting eye colour in *Drosophila*. Their populations were all started with allele frequencies of $p = q = 0.5$. (In nature, the *white* allele is found at negligible frequency as *w*/*w* homozygotes are at a considerable selective disadvantage.)

Jones and Probert investigated the effect of habitat selection on this artificial polymorphism in the following way. They set up three types of cage, each with two compartments between which flies could migrate through a small slit in the partition. In one kind of cage, both compartments were lit with bright white light; in another they were lit with dull red light; in the third, one compartment was subjected to each form of lighting.

It is well known that the white-eyed phenotype is much less fit than the wild-type under normal (i.e. white) lighting conditions—see Figure 3.2 which provides data on this for *D. melanogaster*. It is also less fit than the wild-type in dull red light, though here the selective differential is less pronounced, as this kind of lighting is optimal for the light-sensitive white-eyed flies but suboptimal for the wild-type. Thus given either light regime on its own, the *white* allele declines in frequency and eventually disappears. The question Jones and Probert were interested in was whether, in the 'heterogeneous' type of cage permitting habitat selection, the polymorphism was able to persist.

Data from the three types of population cage are given in Figure 6.1. It can readily be seen that the tendency for the *white* allele to be eliminated is considerably less strong in the cages permitting habitat selection. What is not clear is whether, in these cages, the mutant gene is merely being eliminated more slowly or whether there is a stable equilibrium frequency of *w* somewhere between 0 and 0.5. Related to this, it is not clear whether the system in the heterogeneous cages is frequency dependent. Although there are considerable difficulties in testing for frequency dependence in time-series data, there are ways of getting around these (see Cook, 1983; Arthur and Middlecote, 1984a), and it is a pity that Jones and Probert made no attempt to determine whether the *w* allele was indeed being actively maintained in the heterogeneous cages.

While the experiments of Jones and Probert (1980) certainly do show habitat selection (regardless of whether it actually promotes an equilibrium), the system is one in which the mutant phenotype is at a selective disadvantage in both 'niches'. Thus it does not correspond to the sort of system originally envisaged by Levene (1953) where even in the absence of habitat selection, polymorphism could be maintained if different genotypes were fitter on different resources.

Turning to a 'natural' polymorphism—that of the *Adh* locus producing the enzyme alcohol dehydrogenase in *Drosophila melanogaster*—a study by Day *et al*. (1974) revealed differences in the activities of the 'fast' and 'slow' allozymes which might be associated with niche differences. These authors measured the relative activities of enzyme extracts from the three *Adh* genotypes on a variety of different substrates (ethanol, *n*-propanol, isopropanol, *n*-butanol, isobutanol and cyclohexanol). They found significant variation in relative activities among substrates, though enzyme from the *F*/*F* genotype was always the most active. If the activities measured are directly

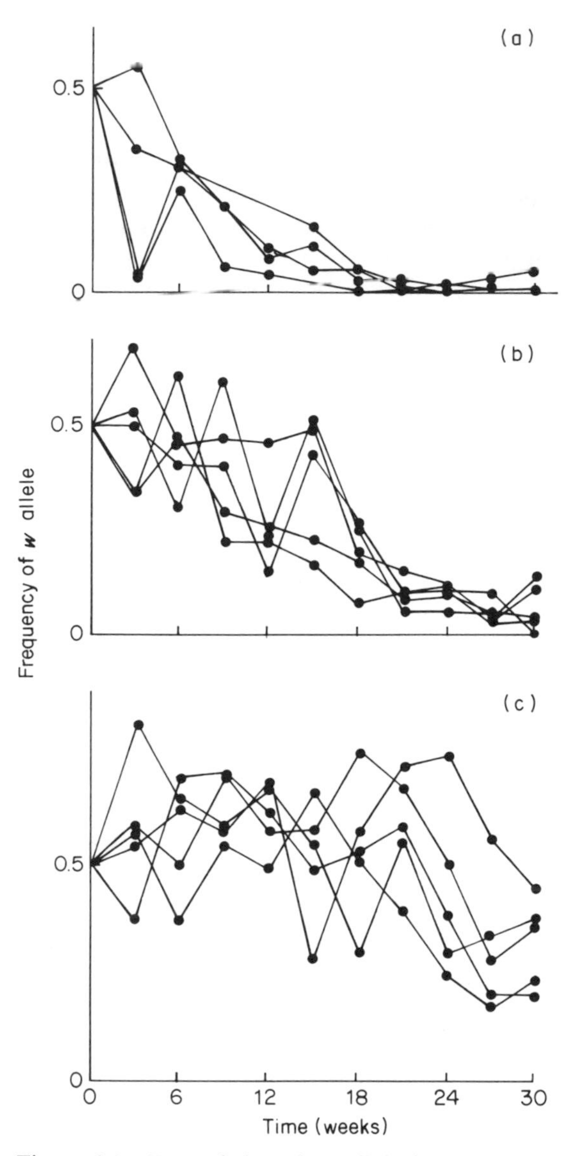

Figure 6.1 Fate of the *white* allele in *Drosophila simulans* in: (a) Cages with white light. (b) Cages with red light. (c) Cages permitting habitat selection. From Jones and Probert (1980) reprinted by permission from *Nature*, **287**, 632–633 © 1980 Macmillan Journals Ltd

related to fitness (which seems unlikely), the fitnesses would vary in a manner similar to that of the eye colour phenotypes in Jones and Probert's (1980) experiment. That is, the 'slow' allozyme would be at a marked disadvantage on some substrates (e.g. isopropanol) but at a smaller disadvantage on others (e.g. *n*-propanol). However, given a more complex relationship between enzyme activity and fitness, it is not inconceivable that the significant variation in activity among substrates demonstrated by Day *et al*. (1974) and also by Vigue and Johnson (1973) could lead to fitness/niche differences of the sort that can maintain a polymorphism in Levene's (1953) model. Whether this is actually the case, however, is not known; nor do we know whether *Adh* genotypes exhibit habitat selection. (For a review of work on the *Adh* polymorphism, see Van Delden, 1984.)

Many other experiments aimed at examining the stabilizing effect, on polymorphism, of niche differences among competing genotypes have been reviewed recently by Hedrick (1986). His conclusion, with which I agree, is that most individual case studies are beset with problems, though some present a stronger case for multiple-niche polymorphism than others; and that the general importance of niche differences as a stabilizing mechanism in intergenotypic competition remains uncertain.

6.3 FIELD STUDIES

As in laboratory work on multiple niche polymorphism, it is necessary in field studies on this phenomenon to make the distinction between differential *choice* of resources/microhabitats and differential *fitnesses* of genotypes in them. One simple way to concentrate on fitness differences is to work on terrestrial plants. Since neither seedlings nor mature plants can migrate to select an appropriate microhabitat, and since none of the various means of seed dispersal can be thought of as involving the 'choice' of a site that is ecologically appropriate for the genotype concerned, it would seem that we can fairly safely rule out habitat selection in the case of plants, with the possible exception of those reproducing vegetatively.

Although they probably do not exhibit habitat selection, plants do have characteristics which could nevertheless lead to multiple-niche polymorphism purely through variation in relative fitnesses between microhabitats. These characteristics include dependence on edaphic factors which can be highly heterogeneous on a small-scale basis, and fairly restricted gene flow. Because of such characteristics, Ennos (1983) concludes that 'a substantial amount of genetic variation could be maintained in plant populations by multiple-niche selection'.

One example of a plant species in which fitness differentials do appear to vary among habitats is the perennial grass *Spartina patens* (Silander, 1979). In a coastal area of North Carolina, this species was found occupying three fairly distinct habitat-types within a small area (along a 200 m transect). These habitats were dunes, swales (moist depressions between dunes) and marshes.

From an analysis of genetic variation in several enzymes, Silander inferred the existence of a large number of different 'combined genotypes'. Although the dune plants included fifteen genotypes, the swale plants 31 and the marsh plants 46, only three genotypes were found in common between dune and swale, two between dune and marsh, and six between swale and marsh.

Clearly, Silander's (1979) results show a very great amount of differentiation among habitats. However, what is lacking in his study is a comparison of different replicate sites from the same habitat-type separated by a similar distance to that separating sites in different habitats. This is particularly important because when there are so many genotypes there may in general be little overlap between those found in different local areas. However, Silander (1979) does report that reciprocal transplant experiments showed that each genotype was fittest in its particular habitat. Also, Silander and Antonovics (1979) demonstrated that plants from the different habitats grown in a common environment were significantly different in several ways indicating adaptation to the habitat concerned (e.g. dune plants most resistant to drought and salt).

Microgeographic differentiation of gene-frequencies in animals has also been described but its interpretation is more difficult because it may represent the effect of variation in relative fitnesses between sites or habitat selection or both. Since it is possible to study fitness aspects in isolation from habitat selection in plants, it would be nice if we could somehow use animal systems to study habitat selection on its own. Essentially, this means that we should study what animals of different genotypes actually *do*. There are many difficulties involved here, not least of which is that what animals do—including their choice of resource or microhabitat—may change from moment to moment. Hence, apparent differences in behaviour between genotypes found at one point in time may be transient, chance differences not repeatable in subsequent study periods.

One interesting way of getting around this problem has been reported by Jones (1982), working with the visible polymorphisms of the snail *Cepaea nemoralis*. Although the causes of variation in morph-frequency from place to place are reasonably well understood, the mechanism of maintenance of the *Cepaea* polymorphisms is still unknown (see Jones *et al*., 1977; Clarke *et al*., 1978 for reviews). One possibility, of course, is that snails of different shell phenotype select different resources/microhabitats, giving rise to a multiple-niche situation. Jones (1982) describes a method of detecting behavioural differences between phenotypes that is based on applying to each shell a dab of a light-sensitive paint. Marked snails recaptured after a period of time in the wild reveal their extent of exposure to sunlight by the degree of paint fading, which is measured against a set of standard shades.

At one level, the results of this study were very conclusive: paint marks on banded shells faded more rapidly than those on unbanded shells in all of a series of ten experimental populations (housed in cages in the field). This difference between the phenotypes was significant within many of the populations and on an overall basis. It seems certain, then, that the banded snails used in this study were exposed to more sunlight, over the 60-day period of the experiment, than the unbanded ones.

In other respects, however, Jones' (1982) study was not so conclusive. First, there is the question of whether snails from other areas would show the same effect. (The experimental snails were all from one Spanish valley system.) Second, we do not know *why* the phenotypes were exposed to different amounts of sunlight. It is possible:

1. That bandeds are active for longer periods.
2. That they are active at different periods (less nocturnal).
3. That they occupy more exposed microsites in horizontal or vertical dimensions.
4. That any combination of these effects occurs.

Finally, as noted in Chapter 2, we do not know what limits the population size of *Cepaea* in the field. If availability of suitable microsites is the limiting factor, then Jones' data may indeed be relevant to the maintenance of the polymorphism; but given any other limiting factor, it is probably not relevant.

What is really needed in a study of habitat selection as an agent of maintenance of polymorphism is (a) identification of the limiting factor and (b) description of RUFs of both (or all) phenotypes in relation to that factor. If this could be coupled with experiments in artificial environments some of which do, and some do not, permit the observed difference in RUFs in the wild, then we would be a lot closer to conclusive demonstration of polymorphism through habitat selection than we are at present.

6.4 CONSTANT OR VARIABLE FITNESSES?

In Section 6.1, I dealt with two models in which stable polymorphism was produced as a result of ecological differences between genotypes. In one of these (Levene, 1953), the fitness of each genotype in each niche was a constant (V_i or W_i). In the other model (Clarke and Allendorf, 1979), fitnesses were variable, and more specifically the fitness of a variant was inversely related to its frequency. Although both of these kinds of system can produce stability, it is necessary to be clear about the distinction between them. This point is also made by Ennos (1983) in his review of genetic variation in plant populations.

This issue is particularly complex in 'multiple niche' situations, and it may help to consider for a moment a very simple system—that of a population of a haploid organism with two allelic variants A and B occupying an environment which contains two microhabitats or resources 1 and 2.

A simple method of estimating relative fitnesses of two variants A and B is the cross-product ratio (see Cook, 1971; Arthur, 1984). This is defined as:

$$\mathrm{CPR} = \frac{N'_{\mathrm{A}} N_{\mathrm{B}}}{N'_{\mathrm{B}} N_{\mathrm{A}}} \qquad (6.3)$$

where N measures the number of organisms at the start of an experiment and N' measures the final number (usually after a single generation). The form given above measures the fitness of A relative to B and is generally labelled w_A. This has a value of 1 for equal fitnesses and a value in excess of 1 when A is the fitter variant. Turning the equation upside down would give w_B. That is, for relative fitnesses of A and B, $w_A = 1/w_B$.

In a one-resource environment, the values of each N are unambiguous. However, in an environment with two resources we have to decide whether to measure fitnesses separately in each resource or to try to devise some overall measure of relative fitness for the two-resource environment as a whole.

This choice is crucial, because it may well be that fitnesses measured one way are constant, while those measured the other way are variable. For example, suppose that fitnesses within each resource are fixed—as they are in Levene's (1953) original model—and that there is a degree of habitat selection such that variant A prefers resource 1 and variant B prefers resource 2, but these preferences are statistical rather than absolute. We may further assume that the fixed fitnesses correspond with the habitat selection in that $w_{A1} > 1$ but $w_{B1} < 1$ and conversely for resource 2. Now although fitnesses are fixed within each resource, it is clear that increasing the number of the A variant will cause more depletion of resource 1 than resource 2 and will consequently reduce the survival probability of individuals in that resource. Since these individuals are predominantly A, this means that an increase in the frequency of A in the overall system will decrease the fitness of A in that system—i.e. there is frequency dependence. Yet we have assumed—and it is perfectly reasonable to do so—that fitnesses *within* each resource are *not* frequency-dependent. So whether or not a system exhibits frequency dependence may depend on how we measure fitness. This is equally true of competition between species, where resource partitioning causes overall competitive abilities to be frequency-dependent, but need not cause competitive abilities within any one resource to be so.

PART III

COMPETITION, COMMUNITY STRUCTURE AND EVOLUTION

In this part of the book I address the wider issues to which the niche concept relates—particularly in the realms of the structure of natural communities, the coevolutionary 'fine-tuning' of the characteristics of competing species (which may or may not make an important contribution to community structure), and long-term large-scale evolutionary changes, especially those that appear to be induced by empty niche space. This part of the book is necessarily more speculative than the parts which precede it. We now know quite a lot about the population dynamics of competition in experimental systems of *Paramecium* and *Drosophila*, but much less about the relevance of such systems to an understanding of nature. To a large extent, I try in this part of the book not to give conclusive answers (which are not yet available) but rather to give a clear statement of what the central issues are that we need to address. Even this more limited aim is not easily achieved, as will become apparent as we proceed.

Despite our lack of detailed knowledge of the wider issues, certain messages, however vague, are beginning to appear. Worthy of note among these are the following three. Coevolution of competing species probably occurs, at least in some systems, but it is a rarer, subtler, more heterogeneous, and more difficult-to-detect phenomenon than was originally anticipated. Community structure has no single dominant organizing force, but interspecific competition *is* one of the group of organizing forces that we find. In the long term, major evolutionary shifts seem to be associated, in a variety of ways, with empty niche space and a *lack* of competition.

Chapter 7

Coevolution

7.1 INTRODUCTION

In Chapters 2 and 5, I dealt with the outcome of competition between species whose ecological properties were taken to be fixed. In other words we assumed that, given a particular pair of species in a particular environment, there will be a certain kind and degree of ecological difference between the species, which will influence whether or not they coexist. However, while it is useful to take this stance initially so that the population dynamics of competition can be examined in as simple a situation as possible, it is widely recognized that the ecological properties of competing species are not entirely fixed, and that they may even change as a direct result of the competitive process itself. Such changes—or rather a subset of them, as will be explained below—are the subject of the present chapter.

Unfortunately, the study of changes in the characteristics of competing species has been plagued by an imprecise and ambiguous terminology, which has led to (and probably also been caused by) some rather muddled thinking. The literature is full of terms like niche divergence, habitat shift, character displacement, genetic feedback, niche expansion, character convergence and many others. Some overall system of naming the changes that occur is necessary wherein the relationships between different sorts of changes are apparent.

One problem with all the above-mentioned terms is that they refer to *phenotypic* changes—whether of behavioural or morphological characters—and such changes can be caused in two quite distinct ways. If we find that, in sympatry, the niches of species A and B overlap less than they do in allopatry, then this may be due to divergent evolutionary changes having occurred in the sympatric area in response to competition. However, it may simply be a

Table 7.1 Heritability values for some morphological and non-morphological characters. From Arthur (1984)

Character	Species	Heritability
Morphological		
Shell diameter	*Partula suturalis*	0.53
Shell diameter	*Arianta arbustorum*	0.70
Beak length	*Geospiza fortis*	0.97
Height	*Homo sapiens*	0.65
Abdominal bristle no.	*Drosophila melanogaster*	0.52
Non-morphological		
Phototaxis	*Drosophila pseudoobscura*	0.10
Egg production	*Drosophila melanogaster*	0.20
Milk yield	Cattle	0.35

consequence of the organisms concerned responding behaviourally to the competition and choosing to utilize different resources to those that they would have utilized in the absence of interspecific competition. It may even be a mixture of the two.

Because behavioural characters tend to have lower heritabilities than morphological ones (see Table 7.1), it makes some sense for the would-be student of coevolutionary changes to concentrate on morphology, and I will do precisely that in the following pages. However, it must be stressed that the distinction between behavioural and morphological characters cannot be equated with the distinction between ecophenotypic and genetically-based ones. Behaviour does, after all, evolve. Also, purely ecophenotypic changes in body size and size-correlated morphological measurements are well known, particularly in response to varying density (see Bakker, 1961, for the classic work on *D. melanogaster*). So although it makes sense to concentrate on morphological characters, this by no means removes the need for evidence that observed phenotypic changes have some genetic basis. Indeed, as we will see, this need is very great if it is being claimed that coevolutionary changes have taken place.

With regard to the plethora of names describing the phenotypic changes that are sometimes observed in sympatry—some of which were listed above—one system of ordering them is as follows. We use 'niche' as a prefix where the changes are behavioural/ecological rather than morphological; in the latter case 'character' should be used instead. Where the behavioural/ecological change takes the form of altered spatial distribution rather than resource choice, then 'habitat' is a more restricted prefix that can be used.

Within any one species or population, the distribution of a character—whether morphological or behavioural—can be summarized by its mean and variance, both of which are subject to change. Variance can increase (*expan-*

sion) or decrease (*compression*), and we need no further terminology than this because it does not seem to make sense to relate the kind of change in variance in one species' niche to the variance in the other. However, we do make precisely this kind of comparison with means. In one species considered alone the mean can *shift* (upwards or downwards on whatever axis is being used); if both species shift, then there may either be *convergence* or *divergence/displacement*. (I use divergence and displacement synonymously like most students of coevolution but unlike Grant, 1972.)

We now have an overall system of nomenclature for coevolutionary changes which has a measure of 'internal logic' and which also reflects past usage to a reasonable degree. For example, the 'classic' cases of character displacement in *Sitta* (Brown and Wilson, 1956) and *Hydrobia* (Fenchel, 1975) are indeed character displacement because they involve divergence of the mean values of morphological characters. Equally, increased variance of perch height and prey size in an *Anolis* lizard population is indeed 'niche expansion' (see Lister, 1976a,b).

Our system is not quite exhaustive because it excludes some proposed forms of coevolution, particularly those, like genetic feedback (Pimentel *et al*., 1965), which involve evolution of 'competitive ability'. Observed changes in competitive ability are a *result*, expressed in terms of changing relative densities, of other *unknown* changes in the means and/or variances of unspecified characters. However, I am going to concentrate in subsequent sections on coevolutionary changes that relate fairly readily to niche theory, so I will not look any further at genetic feedback and related processes. (These are reviewed in Arthur, 1982a.)

In fact, I intend to concentrate to a large extent on character displacement. This seems to have captured the imagination of evolutionary ecologists more than other proposed forms of coevolution, and there are good reasons for this to be so. If species have substantially overlapping RUFs, and if some of the variation that constitutes each RUF is due to genetic differences *between* individuals (Roughgarden's, 1972, BPC), then displacement seems the obvious coevolutionary outcome. Whatever balance of selective forces acted upon the RUFs in allopatry, the move to sympatry adds a new selective pressure against individuals in each species that are most like the other—hence RUFs diverge.

I suspect that it is this 'logical appeal' of displacement that has caused so many ecologists to attempt to find it. Studies revealing displacement in morphological characters are more persuasive than others, because, as we have seen, heritabilities tend to be higher, and it seems less likely that we are dealing with a purely ecophenotypic phenomenon. Studies purporting to show character displacement have been extensively reviewed by Grant (1972) and Arthur (1982a), and there thus seems no reason to list all the purported cases of displacement herein. Rather I will concentrate, in the next section, on making a few general points on character displacement and on discussing some recent case studies which are not yet very widely known.

7.2 CHARACTER DISPLACEMENT

7.2.1 Schematic view and basic theory

Suppose that we have two species whose distributions partially overlap so that there are pure (allopatric) populations of each species as well as mixed (sympatric) populations. Further, suppose that there is a morphological character which in some way reflects the feeding ecology of the organisms concerned (e.g. beak size in birds), and that, in general, the distributions of this character in the two species overlap considerably. This is the background against which character displacement is usually seen as happening, and it is clearly not an unreasonable state of affairs. Indeed, the majority of pairs of closely related species will show the partially-overlapping geographical and character distributions suggested above.

The basic idea of character displacement is simply that, whatever degree of separation of character distributions is found when allopatric populations of the two species are compared, this will be increased in sympatry because selection will favour those individuals in each species in the tail of the distribution that is not overlapped by the alternative species. That is, the bigger species will get bigger and the smaller one smaller with respect to the character being monitored. This process is illustrated in Figure 7.1.

If we imagine a transect of samples extending from an allopatric population of species 1 through a sympatric population into an allopatric population of species 2, then the pattern of change in the mean value of the character measured will be as shown in Figure 7.2. This picture, and that shown in Figure 7.1, together constitute a general, 'schematic view of character displacement; and the situation they describe, as with the 'background' to character displacement given earlier, does not seem too unreasonable. However, unlike the 'background', which has few assumptions and which is frequently found in nature, the apparently reasonable idea of character displacement does have a number of inbuilt assumptions which may affect

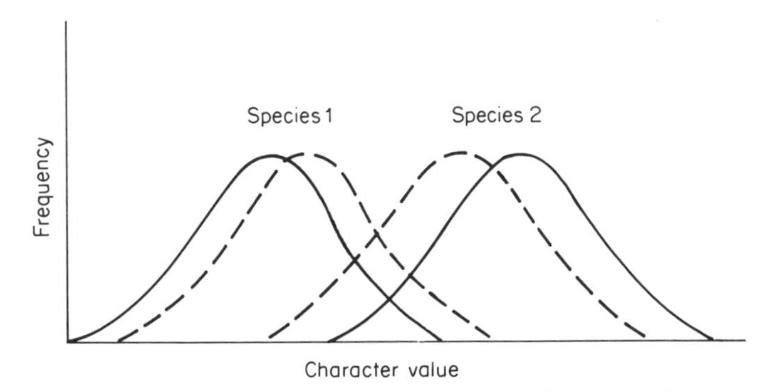

Figure 7.1 Schematic view of character displacement. Dashed lines—distributions prior to selection. Solid lines—distributions after selection

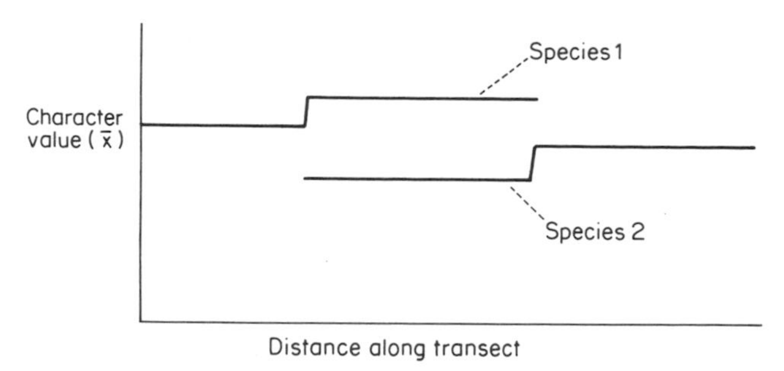

Figure 7.2 Idealized view of character displacement as seen along an allopatry–sympatry–allopatry transect

whether we would really expect the process to be common. It is worth briefly examining what these assumptions are.

First, it is assumed that there is some correspondence between the character distribution and the RUF. There may well be such a correspondence but it is unlikely to be simple. Most RUFs have a large within-phenotype component (WPC); the variation in character values that constitutes a distribution (e.g. of beak size) is entirely variation between phenotypes. That is, each phenotype is represented by a single point in the character-distribution but by a set of values within the RUF. Studies attempting to link morphological and ecological characters have rarely been sufficiently detailed to test this sort of idea.

Further assumptions embedded in our basic picture of character displacement are that the variation is at least partially heritable and that in sympatry there is exploitative competition between the species—that is, they are resource limited and their RUFs overlap. Without competition the form of divergent selection envisaged will not take place; and without heritability the characters involved cannot respond to the selection. Also, it really needs to be the case that RUFs are sufficiently divergent at the start of the process for the species to coexist—because only when there is a fairly extensive and prolonged coexistence will each species be a significant factor in the evolution of the other.

In addition to the above, we need to assume two final things if character displacement is to become a reality. First, there must not be some constraint or counter-balancing selective agent preventing each character distribution moving away from the other. Second, gene flow between sympatric and allopatric populations must be slight—otherwise the effect of the selective process will be diluted or annulled. This will be a particularly severe problem if there is a patchwork of closely spaced sympatric and allopatric populations.

One of the reasons for making the assumptions underlying character displacement explicit at the outset is that when we enter the realm of models it is sometimes hard to see exactly what assumptions have been made. If we know what assumptions to look for it is easier to find them!

In fact, models of character displacement routinely include most or all of the assumptions listed above. One example is the model by Crozier (1974). Crozier considered an environment with five microhabitats/resources and two species competing for them. Species A had three genotypes (AA, Aa, aa) as had species B (BB, Bb, bb), and each genotype of each species had a particular 'utilization phenotype'—that is, it had a particular one of the five resources which was optimal for it, though all genotypes could survive on all resources.

Crozier (1974) devised equations for the fitnesses of the genotypes which had density-independent and density-dependent components as follows. First, the fitness of a particular genotype on a particular resource was specified in a density-independent way, being related only to the 'distance' between that genotype's optimal resource and the actual resource being considered. Thus the fitness of genotype *i* on resource *j*:

$$\mathrm{F}_{ij} = 1 - \frac{|\mathrm{V}_i - \mathrm{V}_j|}{X} \tag{7.1}$$

where V_i is the 'utilization phenotype' and V_j is the optimal phenotype for resource *j*. (X is just a constant.) The 'survival' S_{ij} of each genotype in each resource (really its overall fitness on that resource), was taken to be the product of the density-independent component given above and a density-dependent component, as shown below:

$$S_{ij} = F_{ij}\left(1 - \frac{A_j + B_j}{K_j}\right) \tag{7.2}$$

where A_j and B_j are the numbers of eggs laid by species A and B on the *j*th resource, with K_j measuring the abundance of that resource.

Crozier (1974) used these equations in a computer simulation of coevolutionary change, and found that, in all the 'runs' he presented, character displacement took place—with the proviso that it was in phenotype frequency rather than mean value, since this system involves discrete rather than continuous variation. However, in a way this is a prediction of the inevitable, because the model incorporated *all* of the assumptions given earlier. Heritability and the morphology/ecology relationship were jointly assumed at the outset by equating genotype, utilization phenotype and morphotype. Interspecific competition is assumed in that each species has all its genotypic survivals decreased by the eggs of the other species. Crozier admits that certain numerical values of the parameters were chosen both to permit coexistence and to keep gene flow down to a minimum. Finally, there are no restrictions to competitively induced evolution in the form of counteracting kinds of selection stemming from other ecological sources. Given all these assumptions, and the following relationship between phenotypes and resources:

Resource		1	2	3	4	5
Optimal phenotype	species A	AA	Aa	aa		
	species B			BB	Bb	bb

it is hardly surprising that this coevolutionary system gives rise to character displacement.

Subsequently to the work of Crozier (1974), more sophisticated models of character displacement have appeared, including those of Bulmer (1974), Lawlor and Maynard Smith (1976), Roughgarden (1976), Slatkin (1980), and Milligan (1985). Basically, they all tell the same story: given suitable assumptions, character displacement will follow. One thing that field ecologists ought to be looking for, therefore, is evidence that the theoretical assumptions do actually hold in nature. An alternative approach is to demonstrate a pattern of systematic divergence in sympatry and to rule out any explanation of it other than character displacement resulting from competition. If this can be done, then presumably the various assumptions must be true for the system concerned. This approach will be considered in the following section.

7.2.2 Conclusive demonstration

In the middle of discussing something completely unrelated to coevolution, Whitehouse (1973) makes an important general point that most certainly *is* relevant to coevolutionary studies. The point is that it is most confusing to call a *process* and a related *phenomenon* by the same name. In studies of character displacement we have an even worse situation than this because one process and two phenomena (at least) are all referred to by the 'character displacement' label. I should state at the outset that I use the label herein to mean 'ecological' or 'competitive' as opposed to 'reproductive' character displacement, the latter being a totally separate process (see Brown and Wilson, 1956).

Within the 'competitive' area, character displacement can still mean at least three things. First, the *process* of divergent selection resulting from competition in sympatry, through which sympatric populations become more different than allopatric ones. Second, the *phenomenon* of increased divergence in sympatry, however it was caused. Finally, within sympatry, and in the absence of any comparisons with allopatry, two species are sometimes said to exhibit a certain degree of 'character displacement' in the form of a character's mean value in one species being a certain multiple (often around 1.3 ×) of its mean in the other.

This terminological ambiguity is a recipe for disaster if left untreated, and it makes a lot of sense to dispose of it before proceeding. From here on, I will restrict 'character displacement' to its proper use—namely the *process* of divergent selection in sympatry resulting from interspecific competition and giving rise to a pattern of increased character differences in sympatric populations. I may occasionally also call the resulting pattern character displacement, but only when it is clear that it results from that process. A pattern of variation in which sympatric populations are simply seen to be more different than allopatric ones in the character studied, but it is not clear why they are so, will be called an 'increased difference in sympatry' (IDS) pattern. Sympatric character differences on their own, such as a bird species' beak length being 1.3 × that of its congener, will simply be referred to as *character differences*. Use of this form of labelling should, I hope, preclude the problem of inferring a certain cause for an observed pattern simply because they both have the same name.

Having now disposed of the terminological confusion, we can state the central issue quite clearly: how can we be sure that an observed IDS pattern has been produced by (competitive) character displacement? This overall question can be decomposed into four simpler ones, as follows.

7.2.2.1 Is the IDS pattern real?

The simplest sort of study that can reveal a (bilateral) IDS pattern consists of three samples: one for each species from allopatry and one sample (containing both species) from sympatry. When the character concerned has been measured in this set of samples, we could, conceivably, find a pattern similar to that in Figure 7.1. The problem with studies of this kind is that they tell us nothing about place-to-place variation in the mean value of our character within either allopatry or sympatry. If significant differences among means from different localities are the norm—as is often the case—then the probability of getting an IDS pattern by chance in such a study is 0.25. Despite Grant's (1972) warning that we need many samples to obtain a reasonable picture of variation within allopatry and sympatry before claiming that there is a real difference between them, some subsequent workers have proposed character displacement or other forms of coevolution on the basis of the very simple sort of study described above (e.g. Murphy, 1976).

7.2.2.2 Is the IDS pattern general?

Even if there is a real divergence on entering an area of sympatry along one particular transect, this could be a 'one-off' phenomenon with some special local explanation. Before we conclude that there is something important that requires a general explanation, we need some evidence that the pattern itself is general. Transects into the area of sympatry from other directions will help in this respect. More convincing still would be evidence of an IDS pattern characterizing several independent areas of sympatry in different geographical localities. If many sympatric areas are indeed characterized by an increased character difference when each is compared with its neighbouring allopatric populations, then we clearly do have a pattern that is worthy of causal investigation.

7.2.2.3 Are the character shifts heritable?

One reason why ecological geneticists traditionally study systems of discrete variation, i.e. polymorphisms, is that it is usually possible, with such systems, to demonstrate that the variation is entirely heritable. Observed changes from place to place are thus without question evolutionary changes. Evolutionary ecologists, on the other hand, have concentrated largely on systems of continuous variation. While there are some advantages in so doing—not least of which is that this is by far the commonest kind of variation at the phenotypic level—there is also the twin drawback that (a) fewer of the changes observed are heritable, and (b) it is more difficult to ascertain the heritability of any particular character shift.

Lack of information on heritability is the biggest single failing in most claims that have been made for character displacement. Even the otherwise impressive study of *Hydrobia* (Fenchel, 1975) fails in this respect. Boag and Grant (1978) provide an exception to this lack of information: they furnish heritability estimates for several characters in the finch *Geospiza fortis* and attempt to relate these to the IDS pattern characterizing populations of *G. fortis* and *G. fuliginosa* on the Galapagos islands in general (see Lack, 1947). However, there is a problem here. Boag and Grant (1978) use the standard methods of quantitative genetics, which deal with within-population heritability; and their studies were restricted to variation in morphological characters in the (allopatric) population of *G. fortis* on the islet of Daphne. Unfortunately, the heritability of the variation of a character within a particular population may be quite different to the heritability of a difference in character values between populations. It is the latter that is of interest in studies of IDS patterns.

One crude way to examine whether a character shift between two populations is heritable is to perform a reciprocal transplant experiment. Samples transplanted between allopatry and sympatry should retain their original character values one generation later if the allopatry–sympatry difference is an evolutionary one; if it is a purely ecophenotypic difference, values in the transplanted samples will reverse. It must be admitted that such transplant experiments have certain problems associated with them. One is that they do not provide any recognized measure of partial heritability in cases where the character shift turns out to have both genetic and non-genetic components. Another is that in the unusual situation of a parasite affecting character-values, this would cause differences to appear to be heritable (because the parasite would be unwittingly transplanted also) even though they were not. However, despite these problems, transplant experiments provide a simple and usually effective method of testing whether a character difference between allopatric and sympatric populations is heritable, and it is a shame that would-be students of character displacement have not employed such experiments to strengthen their case.

Perhaps the lack of use of transplant experiments by evolutionary ecologists stems from a feeling that while a particular character shift might indeed represent an isolated ecophenotypic effect, an overall IDS pattern almost necessarily implies a genetically-based process. Nothing could be further from the truth. It is easy to imagine scenarios where purely ecophenotypic influences produce an IDS pattern and therefore, in a sense, 'mimic' character displacement. For example, one of the most pronounced and widespread of ecophenotypic effects is a negative relationship between body size (or size-correlated characters) and density. This phenomenon, which is almost universal in the living world, is discussed in more detail in Chapter 8. The studies that have so far been done related size to density in single-species contexts. It is less clear what will be the 'effective density' to which body size will respond in a sympatric situation. Each species may respond to its own density N_i, the total density N_T, or some more complex measure of density in which a member of the alternative species counts as some unit other than just 0 or 1. One possibility, if competition is asymmetric in the way suggested by Wilson (1975), is that the species of larger body size will largely respond to its own density while the smaller species will respond to the overall (i.e. mixed-species) density. If this occurs, and if sympatric zones are characterized by increased overall densities but lower 'individual' densities, then the outcome will be an IDS pattern, but one whose origin is entirely ecophenotypic.

As well as this general danger for those prematurely claiming to have documented a case of character displacement without performing transplant experiments, there are some specific aspects of some case-studies that strongly argue that information on heritability is necessary. For example, Fenchel's (1975) claimed case of character displacement in mud snails (*Hydrobia ulvae* and *H. ventrosa*) involves a population system whose age (in the range 10–150 years) is barely sufficient for the IDS pattern to have appeared through an evolutionary, rather than an ecophenotypic, process.

I have spent some time on the issue of heritability because without information of this kind we cannot convincingly argue that we are dealing with an evolutionary process. Previous studies have not paid nearly enough attention to the need to demonstrate a genetic basis of the character shifts that make up an IDS pattern. If future studies do not take note of this point and provide the missing information, they do not deserve to be taken seriously.

7.2.2.4. Is competition the selective agent?

If an IDS pattern turns out to be the result of heritable character shifts, then there seems little doubt that the pattern is caused by selection. The very existence of a pattern, especially if it is repeated in different geographical areas, argues against an explanation in terms of stochastic processes. However, while we may be able to accept selection as the causal factor, we cannot equally readily accept that the selection must originate in the process of interspecific competition, that is, that competition is the selective agent.

The problem here is simply that areas of sympatry do not occur at random—rather, they will be located in particular places that allow coexistence, perhaps because the environment is more heterogeneous there than in other places. Given a systematic ecological difference between allopatry and sympatry, this itself could be the cause of observed character shifts. That is, we have two causal sequences:

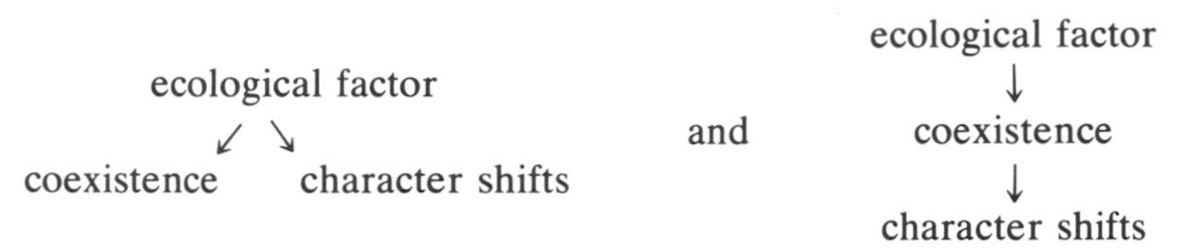

between which we need somehow to distinguish if we are to finally understand the source of an IDS pattern.

Again, the 'alternative' explanation I have constructed is not an unreasonable one. Suppose, for example, that species A is found in hotter, drier areas and species B in cooler, wetter areas, with sympatric populations in areas of intermediate climate. Suppose also that species A is smaller in body size. Now given climatic selection for size within each species on the basis of optimal surface area/volume relationship, size will tend to increase going out of sympatry in species A, and going into it for species B. This, of course, means that an IDS pattern will be generated. Even if there are stepwise changes in character values at allopatry/sympatry borders, we still cannot claim that competition must be the selective agent, because it may well be that those borders tend to be located in the region of most rapid ecological change.

How do we distinguish between the two alternative causal sequences that have been identified? Ideally, we would set up selection experiments in the laboratory where sympatry and variation in climatic or other factors can be separated from each other. Regrettably, most organisms have too prolonged generation times for this to work, unless the selective pressures are very great. Failing conclusive experiments of this kind, a variety of other, less conclusive, lines of approach are open. One obvious approach is to seek out allopatry/sympatry transitions in regions whose predominant habitat-types are as different as possible. If all such regions exhibit an IDS pattern, the case for competition being the selective agent is enhanced. If competition between the species concerned can actually be experimentally demonstrated in the areas of sympatry, then again the case for character displacement is strengthened—though of course there is the problem here that after character displacement the 'strength' of competition will be reduced, and hence it may be difficult to detect. Finally, if the relevance of the character monitored to the competitive process can be established, and better still the morphological/ecological relationship quantified, so that we can see both why and how competitive selection should operate, then the case is stronger still.

7.2.3 Case-studies

The establishment that an observed IDS pattern is real, general and heritable, and the discrimination of competition from other selective agents, are not easy tasks. It is thus hardly surprising that most, indeed all, previous studies of IDS patterns cannot be regarded as conclusively attributing the observed patterns to the coevolutionary process of character displacement. The original 'classic case' of character displacement in nuthatches of the genus *Sitta* (Brown and Wilson, 1956) is certainly deficient in many respects (Grant, 1975). Some more recent studies have been of improved design but still cannot be regarded as conclusive, including Fenchel's (1975) study of *Hydrobia* mud snails, mentioned above, and the study of catostomid fishes by Dunham *et al.* (1979). These and other studies have been reviewed by Arthur (1982a).

In this section I will briefly examine two recent studies of IDS patterns and examine the extent to which their claims to have demonstrated character displacement can be accepted.

7.2.3.1 European shrews

Malmquist (1985) studied a number of allopatric and sympatric populations of the pgymy shrew *Sorex minutus* and the (larger) common shrew *S. araneus* in the British Isles and Scandinavia. He took eight different measurements on the skulls of these shrews, all being essentially measures of length or width, and hence, generally, measures of skull size. Further, since these were expressed as 'primary' variables rather than as a proportion of (for example) the overall length of the animal, we may assume that, to some extent, Malmquist's characters are measures of overall body size.

The evidence presented for the existence of a general IDS pattern is fairly impressive, though it must be stressed that the pattern is a unilateral one, with *S. minutus* diverging away from *S. araneus* in sympatry. There was no detectable effect of sympatry on the size of *S. araneus*. The data on *S. minutus* are shown in Figure 7.3. It is noteworthy that the sympatric and allopatric populations are both of diverse geographical origins. Had one set been confined to Scandinavia, the other to the British Isles, or had one been restricted to British and Scandinavian mainlands, the other to small offshore islands, the evidence for a general IDS pattern would have been considerably less persuasive.

The biggest problem with Malmquist's study, as with most others, is the lack of information on heritability. Presumably voles, like diverse other taxa, show a pronounced negative relationship between body size and density for purely ecophenotypic reasons. Malmquist's explanation of the lack of sympatric divergence in *S. araneus* is that mixed populations have much higher densities of this species than of *S. minutus*. While this is a reasonable explanation in the case of character displacement, it may also be used to

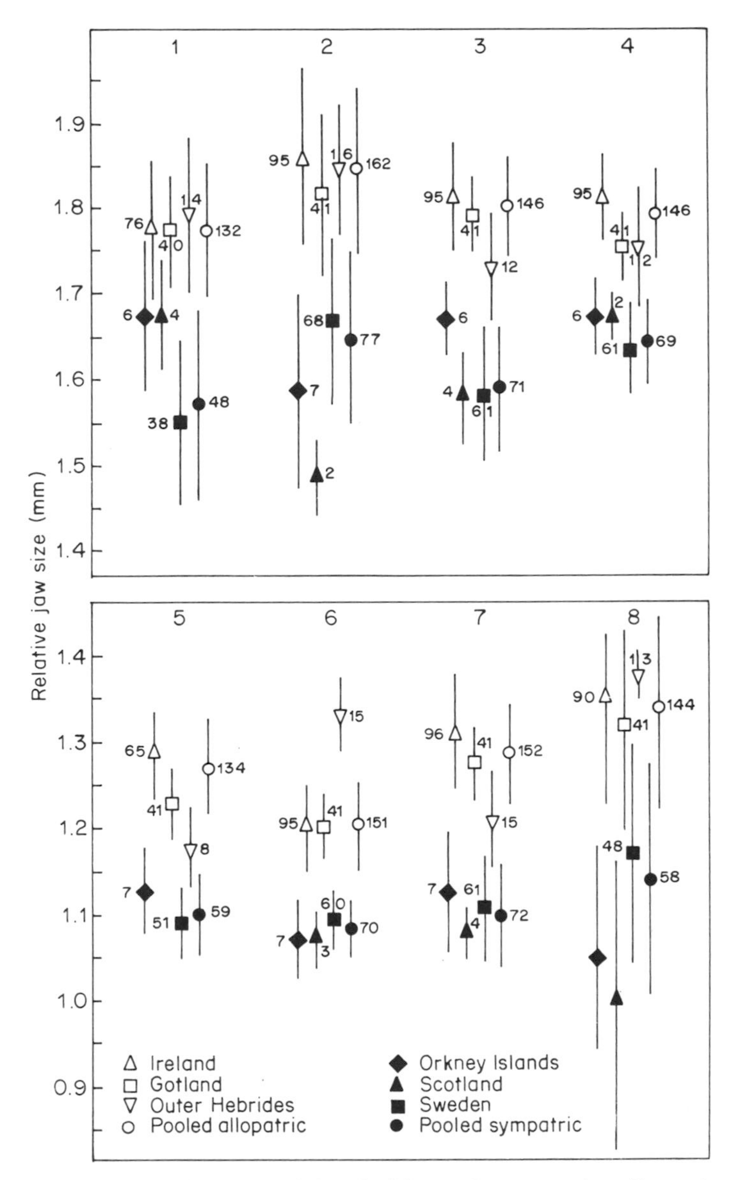

Figure 7.3 Values of eight skull/jaw characters in allopatric (open symbols) and sympatric (solid symbols) populations of *Sorex minutus*. Reproduced by permission of Ecological Society of America from Malmquist (1985) *Ecology*, **66**, 372–377. Bars indicate 95 per cent confidence limits. Numbers close to symbols represent sample sizes

construct an ecophenotypic explanation. *S. minutus* may face a large increase in 'effective density' on entering sympatry, while the transition from allopatry to sympatry may be associated with a relatively small change in effective density for *S. araneus*. This does not, of course, mean that an ecophenotypic explanation is necessarily true for Malmquist's IDS pattern; but it does mean that we cannot convincingly distinguish between this explanation and the evolutionary one that is offered by the author.

7.2.3.2. *Darwin's finches*

One of the earliest IDS patterns to be documented was that involving beak depth in the medium and small ground finches (*Geospiza fortis* and *G. fuliginosa*) on the Galapagos islands (see Lack, 1947). This has frequently been cited as an example of character displacement, with the proviso that the time-sequence in the establishment of the populations studied is thought to be sympatric → allopatric, which is the reverse of the sequence normally envisaged. Although there are some doubts as to whether character displacement is indeed responsible for this IDS pattern (Arthur, 1982a), recent studies have attempted to distinguish more clearly than before between character displacement and some alternative explanations (Schluter *et al*., 1985; see also Schluter and Grant 1984).

The novel aspect of these recent studies is that they quantified the availability of different kinds and sizes of seeds on all the islands involved in the study and used this information to build probability distributions of finches along a beak depth axis. By comparing observed beak-depth distributions with these 'expected' ones, it was possible to examine discrepancies and to see if these were in the directions to be anticipated on the basis of character displacement. Figure 7.4 shows both the expected and observed beak-depth distributions. Two things stand out from the data. First, and most obvious, is that observed mean values (arrowed) correspond quite closely to peaks in the expected distributions, although the comparison is less clear in *G. fuliginosa* because the expected distribution on the left-hand side of the beak-depth axis is messier and tends to be bimodal. Second, the mean values in sympatry (Santa Cruz) correspond less well to their 'expecteds' than do the allopatric values. The discrepancies are in the direction expected on the basis of character displacement, with the data on *G. fortis* again being clearer than that on *G. fuliginosa*.

Although the approach of Schluter *et al*. (1985) is to be commended for its attempts to separate the effects of competition from those of food supply alone, it still cannot be regarded as conclusive. There is no doubt that the IDS pattern is real—these populations have been studied by several workers subsequent to Lack's (1947) pioneering efforts, and the pattern itself is now quite clear. However, its generality is another matter. Unfortunately, all large islands in the Galapagos, with the possible exception of Hood, have sympatric populations (Lack, 1947); allopatric populations are much scarcer and are

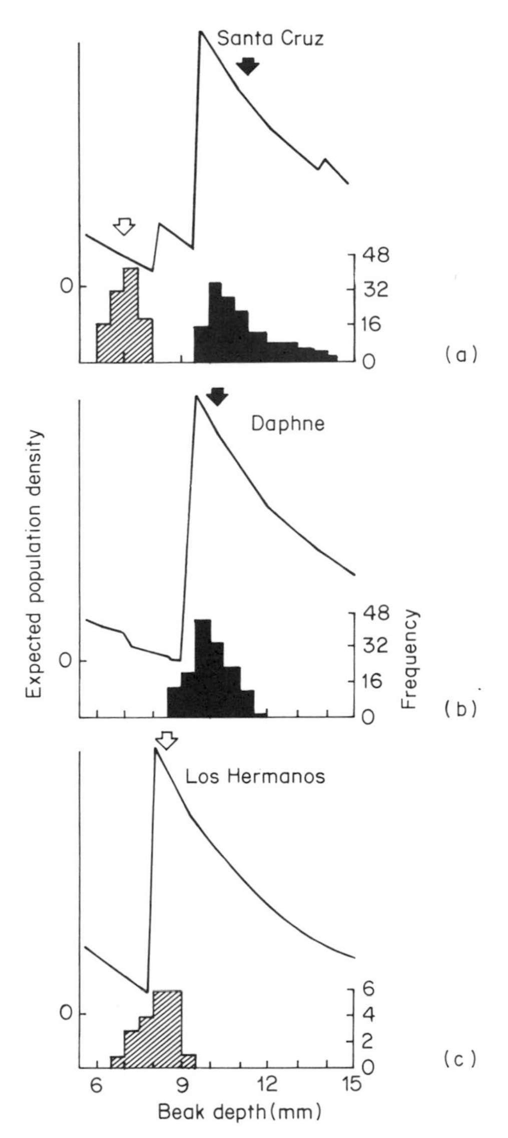

Figure 7.4 Comparison of observed and 'expected' beak depths in *Geospiza fortis* (black) and *G. fuliginosa* (cross-hatched) in (a) sympatry and (b), (c), allopatry. From Schluter *et al.* (1985). Note: Observed—histograms and right-hand y-axis. Expected—solid line and left-hand y-axis. Arrows indicate means of observed distributions. Reproduced with permission from Schluter (1985). Copyright 1985 by the AAAS

restricted to small islets. Exactly how many allopatric populations of *G. fortis* and *G. fuliginosa* exist is rarely stated, and readers of the vast literature on Darwin's finches are left wondering whether there are, for example, any allopatric populations of *G. fortis* other than the one on Daphne for which data are always given. (Apparently there are not—P. R. Grant, personal communication.) This leaves a question mark over the generality of the IDS pattern.

Another problem that remains in this study is that of the heritability of inter-island differences in beak size. As noted earlier, *G. fortis* is one of the few species used in studies of IDS patterns for which heritability estimates are available (Boag and Grant, 1978). However, we still have no estimates of the heritability of the actual differences that make up the IDS pattern, as opposed to estimates of the heritability of within-population variation; and we have no reason to believe that these two heritabilities will be the same.

It is in the discrimination between alternative selective agents, assuming that the observed differences in beak-depth *are* genetically based and selectively driven, that Schluter *et al.* (1985) have made an advance over most other studies. Admittedly, they have only looked at one other factor, the food supply, but this does seem the most sensible one to choose, given the nature of the character being measured. The most reasonable explanation of the data shown in Figure 7.4 would indeed seem to be, as the authors claim, one that incorporates both competitive and food supply effects, rather than just the latter.

Although one can distinguish between these two kinds of effect, it is of course true that competition works through the food supply. If character displacement is to occur, one or both species must have a significant impact on the 'shape' of the resource spectrum, that is, on the relative availabilities of foods at different points along it. If populations are resource-limited, this would be expected to be the case. If, however, populations are limited at a lower level by the density-dependent action of predators or parasites, then their impact will be relatively slight. This question of the 'direction' of population regulation, which is central to a wide range of ecological issues (much broader than, but including, character displacement), will be returned to in Chapter 8.

7.3 OTHER COEVOLUTIONARY PROCESSES

As was mentioned earlier, several competitively-induced forms of coevolution other than character displacement have been proposed. Most of these are subject to similar difficulties to those described in the previous section, particularly that of unknown heritability. However, one group of studies avoids the heritability problem by monitoring polymorphic rather than continuously-variable characters. While discrete variation does not necessarily imply a genetic basis, this is usually the case, and anyhow tests of the inheritance of polymorphic variation are easily conducted—simply by looking for Mendelian ratios in the progeny of crosses between different phenotypes.

Very few case-studies have been conducted on the effect of sympatry on polymorphic variation in the field. I know of only three such studies and coincidentally all involve species of molluscs. Murphy's (1976) study of intertidal limpets of the genus *Acmaea* was deficient in design, as noted earlier, and it is not even clear whether there really is any effect of sympatry on the polymorphism studied. Gosling (1980) found sympatric *convergence* in gene frequencies at the phosphoglucomutase locus in two species of marine cockle of the genus *Cerastoderma* and concluded, probably correctly, that this was due to the effects of a common environment and had nothing to do with competition. The only case study which *may* illustrate a competitive effect on gene frequency is my own study of the shell banding polymorphism in *Cepaea* (Arthur 1978, 1980c, 1982b), but even here we cannot be certain that competition is indeed the selective agent. I will discuss this study in some detail because it illustrates quite well the difficulties faced by all coevolutionary investigations that get around the problem of heritability and come up against the 'final' problem of distinguishing competition from other selective agents such as the climate.

The species involved in this study, *C. nemoralis* and *C. hortensis*, share a polymorphism of shell banding, with absence of bands being dominant to their presence (Cain and Sheppard, 1957; Murray, 1963). Both allopatric and sympatric populations are found and, as we saw in Chapter 2, there is some evidence, albeit circumstantial, that there is competition between the species in sympatry, particularly in sand-dune habitats. If the frequency of unbanded *C. nemoralis* is monitored on transects from allopatry into sympatry, a significant decline is often encountered at the boundary between the two types of population (Arthur, 1978, 1980c). Enough geographically-separate areas show this effect for it to be regarded as fairly general, though it must be stressed that not all areas of sympatry exhibit a decreased frequency of unbandeds. (On the other hand, none of the sympatric areas studied showed a significant increase in frequency.) There is some evidence for a parallel effect in *C. hortensis*, but this is not yet strong enough to warrant a conclusion that sympatry with *C. nemoralis* has any general effect on *C. hortensis*' banding polymorphism.

The most pronounced frequency change on entering sympatry occurs in the only known sympatric population on English sand dunes, which is located between Seaton Sluice and Blyth in Northumberland (Arthur, 1982b). This is shown in Figure 7.5. It must be stressed that, while this decline in the frequency of unbandeds represents a divergence—since the coexisting population of *C. hortensis* has a generally high unbanded frequency—some other declines in other areas represent convergences. That is, while the direction of change within *C. nemoralis* is consistent, it does not appear to be related to the prevailing phenotype frequency in *C. hortensis*.

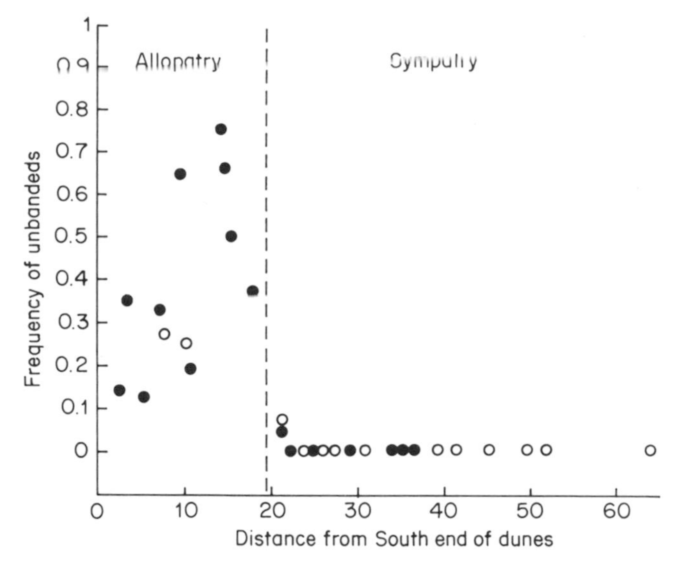

Figure 7.5 Variation in the frequency of the unbanded phenotype of *Cepaea nemoralis* along a transect from allopatry to sympatry in a sand-dune population. Reproduced by permission of *Heredity* from Arthur (1982) *Heredity*, **48**, 407–419. ●—1976 samples; ○—1980 samples. Distance is measured in units of approximately 35 m

One possible explanation of the decline in unbanded *C. nemoralis* on entering sympatry is that unbandeds are weaker interspecific competitors than bandeds. In favour of this explanation are:

1. The consistency in the direction of change.
2. The fact that the most pronounced change occurs in the habitat-type in which competition is thought to be most severe.
3. The precision of the correspondence between the change in frequency and the allopatry/sympatry border (see Figure 7.5).

However, against this explanation are the facts that (a) there is no known reason why unbandeds should be weaker competitors, and (b) there is an alternative selective mechanism which might produce a pattern similar to the one observed.

The alternative mechanism involves climatic selection. It is generally recognized that *C. nemoralis* is found in warmer and drier localities than *C. hortensis*, and that mixed colonies occur in habitats where conditions are intermediate. Also, it is known that banded shells take up heat faster than unbandeds and equilibrate at a higher internal temperature (Heath, 1975). Given these two observations, we would expect the frequency of unbandeds in *C. nemoralis* to increase going into the (warmer) allopatric regions or, conversely, to decline in sympatry—precisely what actually happens.

At present it is impossible to say which of the 'competitive' and 'climatic' hypotheses is correct (Arthur, 1982b). The only truly satisfactory way to separate them would be to perform selection experiments in the laboratory, where climatic changes and sympatry can effectively be separated from each other. While this sort of thing can be done easily enough with *Drosophila*, the generation time of *Cepaea* (about three years) is too long for such experiments to be practicable. Failing this, the most potentially informative study would involve a *large* series of transects from allopatry into sympatry for *C. hortensis*, in which species the two hypotheses predict opposite effects. Unfortunately, the small number of studies that have so far been performed on *C. hortensis* have shown a few weak effects in the direction predicted by the 'competitive' hypothesis, a few cases of no significant change in sympatry, and so far no cases of increased frequency of unbandeds, as predicted by the climatic hypothesis. This rather patchy outcome does not strongly support either hypothesis, and leads to the inevitable conclusion that the selective agent cannot yet be identified.

7.4 COEVOLUTION AND STABILITY

So far, I have treated questions concerning the stability of competitive interactions (Chapter 5) and the coevolutionary effects of these interactions (this chapter) as if they were largely separate matters. Of course, this is not so, because coevolutionary changes may affect the stability characteristics of the system in which they occur and, conversely, whether or not a competitive system is stable has a great effect on the scope for coevolutionary change. This interrelationship is the subject of the present section. Unfortunately, since our knowledge of coevolution itself is so rudimentary, it is impossible to treat the coevolution/stability relationship in much detail; but at least the basic possibilities can be outlined.

The easier of the two questions to address, at least at an elementary level, is the effect of stability on coevolution. Since differentials in competitive ability between species tend to be much greater than selective differentials among genotypes within species, it follows that an unstable competitive system resulting in competitive exclusion will allow little chance for coevolutionary change to occur. Stable coexistence, on the other hand, provides an indefinitely long period of sympatry in which coevolutionary changes may be manifested.

What of the effects of coevolutionary changes on stability? This is a much more difficult question to address, and both stabilizing and de-stabilizing coevolutionary effects may be envisaged. Character displacement is generally taken to have a stabilizing effect. Such a conclusion seems reasonable, at least in systems where some limiting similarity applies. Of course, there is no easy relationship between the amount of character difference that will evolve through character displacement, and the amount required for stable coexistence, as has been noted by Abrams (1983) and others. Indeed, if one accepts that significant coevolutionary change is only likely to take place in the context of an already-stable coexistence, one is assuming that character displacement *starts* with a difference which exceeds the 'limiting' amount. However, given an environment of varying stochasticity, and hence variable limiting similarity, any interspecific divergence could be taken to increase the probability of stable coexistence in the long term.

Forms of coevolution other than character displacement may well *decrease* the probability of stable coexistence. Evolutionary changes in competitive ability provide one example of such a relationship. Evolution of increased competitive ability in *Drosophila hydei*, in competition with *D. melanogaster*, caused the collapse of a state of stable coexistence (Arthur and Middlecote, 1984b)—though the interpretation of this phenomenon is made more difficult by the unexpectedly slight effects that each species had on the other (Arthur, 1986). Where changes in competitive ability occur in both species, this could conceivably lead to a form of stable coexistence, as proposed by Pimentel *et al*. (1965); though there are several reasons for doubting this possibility (Levin, 1971; Arthur, 1982a).

Chapter 8

Interspecific Competition and Community Structure

8.1 INTRODUCTION

We now come to the most important, most difficult, and currently most disputed aspect of interspecific competition. This is the question of whether, to what extent, and in what ways competition influences the structure of natural communities. This is an exceedingly difficult problem—or series of related problems—to address, and there are various pitfalls into which many authors previously addressing it have fallen. It is in relation to the questions of community structure that we encounter most vividly Lewontin's (1974) aptly-named 'agony of community ecology': that is, the agony of not knowing in what terms the system we are investigating will eventually be adequately described and, related to this, the agony of not knowing which questions it is most sensible for us to ask.

Some lessening of this problem is achieved by restricting the discussion to the role of competition in community structure rather than attempting some overall analysis of communities. However, even in this relatively restricted field, the most sensible questions are not readily identifiable; and at any rate, the extent to which competition is an important agent of community structure is often assessed in relative terms (e.g. compared with the extent to which predation is important), so we still cannot entirely omit processes other than competition from our investigation.

Because of the complexity of community structure, it is probably best to begin by eliminating those aspects of it in which competition does not appear to play an important role. The most obvious of these is the number of trophic levels. It is generally recognized that there are no more than five trophic levels in any natural community, though the fact that most 'levels' are not very clear-cut, with some organisms being of mixed trophic-level status, introduces some uncertainty into any such generalizations.

Regardless of the difficulty in delineating trophic levels, it is clear that their number is ultimately restricted either by limitations on the efficiency of energy transfer between them (see e.g. Slobodkin, 1961), or by the 'dynamic instability' of systems with many levels (Pimm and Lawton, 1977). Also, some systems may have fewer trophic levels than would be energetically or dynamically feasible either because evolution has not yet produced the right kind of organism to utilize the current 'top carnivore', or because this organism exists but has not yet migrated into the area concerned.

Since predation, broadly defined, is the means of interaction between trophic levels, while competitive interactions occur within levels, it is clear that a population-based analysis of community structure will need, eventually, to explain trophic-level constraints in terms of predation (regardless of whether the constraints are energetic or dynamic). Aspects of structure within any single level may end up being described in terms of competition (and other non-trophic interactions) though it is clear from the work of Paine (1966) and others that predators can have a marked effect on structure within the level beneath them. Our central question becomes, then: are there any widespread forms of structure within trophic levels, and if so what form do these take? If we can answer this question, then we will at least see the kinds of phenomenon that may (or may not) be explicable in competitive terms.

In fact, even the trophic level may be too large a unit of community structure in which to look for competitive effects. In a deciduous forest, for example, leaf-eating insects of the canopy layer and herb-consuming slugs on the forest floor are unlikely to have any detectable effect on each other, despite the fact that both are herbivores. Rather, interspecific interactions of a competitive nature are likely to be confined to more restricted community units—probably to what is often termed a guild.

The guild was first formulated as an ecological concept by Root (1967) as follows: 'A guild is defined as a group of species that exploit the same class of environmental resources in a similar way. This term groups together species, without regard to taxonomic position, that overlap significantly in their niche requirements. The guild has a position comparable in the classification of exploitation patterns to the genus in phylogenetic schemes. As with the genus in taxonomy, the limits that circumscribe the membership of a guild must be somewhat arbitrary.'

Several points in this definition require clarification. First, a guild is an ecological unit with no necessary taxonomic coherence. This point has been stressed by Jacsic (1981), who points out that particular investigators often wrongly assume that a goup of birds, lizards or whatever correspond to a guild. While I must agree that there is no necessary correspondence between taxonomic and ecological groupings, particularly in the light of studies revealing competition between distantly related forms (e.g. ants and rodents; Brown and Davidson 1977), I suspect that there will usually be such a correspondence. That is, I suspect that Darwin's (1859) often-quoted comment about competition being most severe between closely related species is correct as a probabilistic statement, and that the work of Brown and Davidson (1977) and others is better interpreted as revealing exceptions rather than a new rule.

The second point regards the interpretation of using resources 'in a similar way'. Jacsic (1981) makes the very sensible point that the relevant criterion here is the effect of resource use on the resource itself and its availability for other consumers. Thus if a seed is removed from the overall supply of seeds available for a particular granivorous guild, it does not matter how it was removed or what happens to it after removal. This point is particularly important in relation to some kinds of resource, e.g. leaves, which can be utilized in many different ways. This can be interpreted as a criticism of those who attempt to define feeding niches in terms of how, in behavioural terms, the feeding is accomplished. Different feeding behaviours are only important inasmuch as they cause differences in what part of the resource is being consumed or in the rate of consumption.

The third point is that the existence of guilds does not imply the existence of interspecific competition. This point is crucial, particularly if we are to avoid questions such as 'is competition important in guild structure?' becoming circular. Essentially, membership of a guild implies considerable niche overlap with some or all other members of the guild, but niche overlap is only one of the two necessary and sufficient conditions for exploitative competition. The other condition is that resources are limiting, and the delimitation of a guild is not dependent on, and does not imply, that the resources utilized by that guild do indeed limit the populations concerned. As we shall see, this is really the central question, and we certainly must not assume an answer to it at the outset.

A final point on Root's (1967) definition of the guild is that it probably represents the only realm in which the feeding activities of a group of species may reasonably be viewed in terms of a series of partially overlapping RUFs (on one or many resource axes). This is not to say that guilds are so clearly separated that there is no overlap in feeding between any member of one guild and the members of another. However, it is doubtful if the overlapping-RUF picture is particularly useful in examining these relatively weak inter-guild interactions. Coupled with this point, we might also hypothesize that if competitively-induced coevolution turns out to be an important process (which remains in doubt), then it is likely to be largely restricted to being a within-guild phenomenon.

Having identified the guild as the unit of the community whose structure will most clearly show the effects of interspecific competition if this process is indeed an important agent of community structure, I will now proceed as follows. First, I will examine what is required for a guild to be a competitive one; and what we would expect of its structure, assuming this to be determined by competition (Section 8.2). I will then turn to criticisms of the 'competitive guild concept' and will examine alternative possible guilds in which any observable structure is brought about by other means (Section 8.3). Finally, in Section 8.4, I will examine the extent to which the 'competitive' and 'alternative' views of guild structure are compatible.

8.2 THE COMPETITIVE GUILD

The idea that competition between species is an important agent of community structure, and that the niche concept in its various forms is useful in the analysis of competitively-determined aspects of structure, has a long history, notable landmarks in which are the books of Elton (1927), Lack (1947, 1971), Hutchinson (1965) and MacArthur (1972). During the course of this history, there has been an increasing tendency to interpret community patterns as consequences of interspecific competition without clear evidence that competition was actually occurring in the first place, and without evaluation of alternative possible explanations of the same patterns. This has given rise in recent years to a rebellion against the competitive view of communities, by authors such as Connor and Simberloff (1978, 1979), Strong *et al*. (1979), Pimm (1984) and Lawton (1984).

There are three distinguishable aspects of the competitive view. First, there is the question of whether competition is actually common in natural communities. Second, there is the question of what structure it induces within a guild in which it occurs. Finally, there is the question of what patterns (e.g. of species diversity) emerge from a comparison of *different* competitively structured guilds. I will deal with these three aspects in turn.

8.2.1 Necessary and sufficient conditions for competition

As we have seen in Section 8.1, there are only two conditions required for competition to take place, namely niche overlap and resource limitation. Given these conditions, then competition must occur—that is, the conditions are indeed both necessary and sufficient. It should be stressed that no particular degree of overlap is required for competition—any amount will suffice. (This is quite different from the requirement for *stable coexistence*, where a particular limit to niche overlap may apply—see Chapter 5.) Of course, negligible niche overlap will give rise to a negligible competitive effect, and this may well be impossible to detect; but what is happening and what is detectable are two quite separate issues.

It is hardly necessary to discuss the evidence for niche overlap. The biosphere is simply not populated by species with disjunct niches; and we would hardly expect it to be, since speciation mechanisms, so far as currently understood, will not promote such a state of affairs. Many studies reveal overlapping RUFs in diverse taxa and environments. Examples with which I am familiar include landsnails of the genus *Cepaea* (Carter *et al*., 1979), trichopteran larvae in streams (Townsend and Hildrew, 1979), groups of *Drosophila* species consuming rotting fruits (Atkinson and Shorrocks, 1977), urchins in seagrass meadows (Keller, 1983) and grazing ungulates of African plains (Sinclair, 1985). Countless other examples exist in the ecological literature. Overlapping RUFs must be the norm for congeneric species inhabiting the same environment. Counter-examples could no doubt be found, particularly in very specialist consumers such as insect parasitoids (which tend to be mono- or oligophagous), but these would undoubtedly form a minority, among coexisting congeners, compared with the numerous cases of niche overlap.

The one proviso that must be borne in mind, though, is that for competition to occur RUFs must overlap on *all* relevant axes. Resource segregation on any 'spectrum' describing the limiting resource will prevent competition. Perhaps, on this basis, it would be better to say that 'lack of resource segregation' rather than 'niche overlap' is necessary for competition. This might help to prevent us making the false equation:

'overlapping unidimensional RUFs + resource limitation = competition'.

Turning to the second condition for competition—resource limitation—we find a very different state of affairs. It is most certainly not obvious that limitation by resources is the norm rather than the exception in natural communities. Those of us who have worked on competition experiments in the laboratory may come to expect resource limitation in the field also, simply because we view our model systems as representative in the fundamentals, while lacking much of the detail, of nature. It seems likely that concentration on laboratory experiments has had this sort of effect on many well-known ecologists including Gause (1934) and Nicholson (1933). However, what we are apt to forget is that laboratory experiments in which competition is to be investigated are deliberately engineered to prevent the two major alternatives to resource limitation from taking place. That is, by keeping conditions of temperature, lighting, humidity and so on constant and more or less optimal, we preclude the possibility of having permanently sub-equilibrial populations of the sort envisaged by Andrewartha and Birch (1954; see also Andrewartha and Birch, 1984); and by excluding predators and, as far as possible, parasites, we prevent a form of density dependence that can cause populations to reach a lower equilibrium than that set by their food supply.

What evidence have we that populations in nature are limited by their food? I will consider briefly below the evidence for such a state of affairs; that against will be considered in Section 8.3. Basically, the evidence for food-limitation is of three different sorts: subjective observations (sometimes accompanied by data) that a consumer population is devastating its resource population, and itself being limited as a result; evidence of the well-known signs of food limitation in individual organisms, particularly reduced body size (and the reduced natality that goes with it); and the results of manipulative experiments conducted in the field.

Dempster (1982) reports a long series of investigations of the cinnabar moth *Tyria jacobaeae*, which feeds largely on ragwort (*Senecio jacobaea*). Throughout his paper, Dempster classifies years into those in which the moth defoliated its food plant and those in which it did not. He attempts to show that this has a significant effect on the moth population. While such claims on their own should be regarded with some cynicism, Dempster also provides a considerable amount of data to back up his view that *T. jacobaeae* populations are food-limited. Among other things, he shows that after a 'defoliation' year, when density of moths per unit resource is high, pupal sizes are significantly reduced. This is almost certainly a sign of food limitation, and its corollaries are reduced adult size and reduced fecundity.

In fact, a negative relationship between body size and density seems to be a very common phenomenon in nature; and it provides some of the strongest evidence for food limitation. The classic work on food limitation causing a negative size/density relationship was, as noted earlier, that of Bakker (1961) on laboratory populations of *Drosophila melanogaster*. However, since Bakker's pioneering study, a vast literature on the same phenomenon occurring in *natural* populations has accumulated. The range of examples has spread to include most major taxonomic groups, including insects (Pearson and Knisley, 1985; Grimaldi and Jaenike, 1984), molluscs (Williamson *et al*., 1976), vertebrates (Campbell, 1971; Bobek, 1971) and plants (Westoby, 1984).

Since the variation in size that we are dealing with appears to be an ecophenotypic phenomenon, it is difficult to see how the patterns observed could be compatible with the ideas of either permanently sub-equilibrial or predator-limited populations. (Had it been a genetic phenomenon, one could perhaps have speculated about predators concentrating on dense populations and selecting for small size by preferentially consuming larger individuals.) However, there are two problems which prevent us from using the size/density relationship as conclusive evidence of food limitation in natural populations. First, there is the problem that in many studies (e.g. Williamson *et al*., 1976) density is measured as number of individuals per unit area, not per unit resource. If each local population equilibrates at a certain number per unit resource, and resource levels vary from place to place as they inevitably will, then different densities per unit area should reflect different abundances of resource but the *same* level of intraspecific competition. Why, then, should we have a negative relationship between density and size? One possible answer is that natural populations are rarely if ever exactly at their equilibrium value (which itself will keep changing anyway); rather, at any moment of time, they will either have been perturbed below it (e.g. by an adverse climate), or will have 'overshot' it. Since populations with high densities in terms of numbers per unit area will contain more cases of 'overshot' populations and fewer cases of 'downward-perturbed' ones than populations at low density, we could still expect to get our negative size/density association as a consequence of food limitation, but the argument is more contorted than it first seemed.

The second reason why we cannot conclude that a negative size/density association results from limited food is that there is an alternative explanation—namely that the associations are caused by 'interference' effects. That is, small size at high density is due to a direct antagonistic effect of some sort—chemical or physical—emanating from the other individuals in the population. In favour of this explanation is that it ties in more easily with densities measured as numbers per unit area. Against it is that different mechanisms would be required in different taxa, so that the explanation of a general phenomenon becomes very piecemeal. For example, Williamson *et al*. (1976) who support the 'interference' hypothesis in their study of populations of land snails, propose inhibitory effects of mucus trails as the mechanism. Clearly, this could not be extended to other groups such as vertebrates or plants, where one then has to propose (say) aggression and excretion of harmful products into the soil as mechanisms.

At present it seems clear that a negative association between body size and density in natural populations is evidence for intraspecific competition; but whether it is competition for limited food or competition by interference usually remains obscure.

The alternative to testing whether the necessary and sufficient conditions for interspecific competition are fulfilled is to experimentally test for competition itself, that is, to perform manipulative field experiments designed to demonstrate a (–, –) interaction between the species concerned. I will briefly discuss the design of such experiments, and then turn to an examination of their results.

It should be stressed that the ideal kind of experiment is a reciprocal explant, where three types of population are monitored—an unmodified mixed population, an originally mixed population in which species A has been removed or drastically depleted, and its counterpart involving removal of B. If, in a series of replicates of this triple treatment, ΔN for each species measured over an appropriate period is significantly greater in the artificially-allopatric sites than in the controls, then competition has been demonstrated. Although a reciprocal implant experiment could also show competition in a certain sense, it would really only prove that competition can be induced in natural populations, which is no great surprise and not a tremendous advance over experiments in the laboratory. The successful explant experiment, on the other hand, proves that competition was actually occurring prior to experimental manipulation, which is a much more informative result.

One qualification that must be made in relation to the above design is that it cannot distinguish exploitative competition between species from interference or Holt's (1977) apparent competition. In order to be sure that a (–, –) effect is achieved through use of a common limiting resource, it would be necessary to perform a separate experiment in which resource levels, rather than the consumer populations, were manipulated. Using the same logic as above, resource addition experiments are preferable to those involving resource removal, as the latter only show that exploitative competition can be induced.

Although it was possible to claim, only fifteen years ago, that very few field experiments aimed at demonstrating interspecific competition had been conducted (Williamson, 1972), the situation has now changed dramatically. Hundreds of experiments, on a very wide range of taxa, have been carried out. Since it is impractical to discuss these individually here, I will concentrate on two recent reviews of this mass of experiments—those of Shoener (1983) and Connell (1983). Using slightly different criteria to delimit what constitutes a field experiment on interspecific competition, Schoener puts the number of experiments known to him as 'over 150' and Connell at '72 papers including 215 species with 527 field experiments'.

The criteria used by both authors for inclusion of a study in their review are generally encouraging. Both, for example, exclude studies without adequate controls. Both authors exclude experiments conducted in the laboratory, but Schoener (1983, p. 242) stresses the difficulty of deciding where the laboratory stops and 'the field' starts, and his description of the continuum from one to the other should, in my opinion, be compulsory reading for all ecology students!

Although there are some differences in the conclusions reached by Schoener (1983) and Connell (1983), which have been commented upon by Schoener (1985), there are also many similarities, as follows. First, both authors found evidence for competition in the majority of the studies they examined, and both claim to have found competitive effects in the majority of *species* involved. (Actually, if one defines competition as a (−, −) interaction only a species pair, not a species, can show competition!) Both authors found temporal and spatial variation in competition, though Schoener states that this was largely confined to the intensity of competition, while Connell extends this conclusion to its incidence. Both authors found a considerable amount of strongly asymmetrical competition. Finally, Schoener concluded that exploitative and interference mechanisms were responsible 'about equally often' while Connell draws no general conclusion on their relative importance.

Clearly on the basis of these very extensive reviews, interspecific competition cannot be written off, as some recent authors have done, as a rare and unimportant interaction in natural communities. There certainly are competitive guilds in nature, and a sizeable number of them as well. Of course, other questions remain open, in particular:

1. what overall *proportion* of guilds are competitive (it may yet be a minority); and
2. whether it is possible to generalize about 'where' competitive guilds will tend to be found (e.g. in certain trophic levels, latitudes or taxa).

These questions, and particularly the latter, which is more meaningful, will be discussed in Section 8.4. The remainder of the present section is devoted to an examination of patterns of structure within and among those guilds that are competitive.

8.2.2 Within-guild structure

How would we expect a competitive guild to be structured? The answer to this question depends on which version of competition theory we accept. What follows is based on the assumption that the 'mechanistic competitive exclusion principle' of Chapter 5 is true; and that, associated with this, some degree of limiting similarity—though probably a variable degree—applies in natural communities. Thus I am dealing, in a sense, with the 'niche theory' version of 'competition theory'! It must be borne in mind, though, that mechanisms of competitive coexistence other than those based on classical niche differentiation exist (see Section 4.4), and that a body of competition theory based on any of these would make very different predictions for guild structure. For example, guilds of insect species utilizing a resource distributed in temporary patches, and coexisting because of 'chance' spatial aggregation (see Section 4.5) may exhibit tighter species-packing than conventional niche theory would permit (Atkinson and Shorrocks, 1984). So if it turns out that certain guilds are not structured as described below, this does not automatically lead to the conclusion that they are non-competitive guilds; an alternative interpretation would be that they are competitive ones to which our currently most developed body of competition theory does not apply.

The simplest structure that could be postulated on the basis of 'niche theory' is that of regularly dispersed RUFs on some axis describing the limiting resource. This is shown in Figure 8.1, which also shows the regular dispersion of beak-size distributions for Darwin's ground finches, which has often been interpreted as an example of such a structure (data from Lack, 1947). This kind of proposed structuring has come in for much criticism in recent years. It is thus worth examining in some detail the basis on which it would be expected. A pattern of regular dispersion of niches would seem to require the following:

1. The resource spectrum is infinitely long, or if not then there are no 'end effects'.
2. There is free migration into the area concerned of species that can make use of all sections of the spectrum.
3. All species have the same size and shape of RUF on the dimension depicted.
4. Other dimensions are irrelevant to the spacing of RUFs on this one.
5. Resource is equally plentiful at all points along the spectrum, thus creating no tendency for RUFs to 'clump'.
6. A limiting similarity criterion for stable coexistence applies, and this is independent of how far along the resource spectrum the species-pair we are dealing with is found.
7. Species migrate into the area in an order that relates to their position on

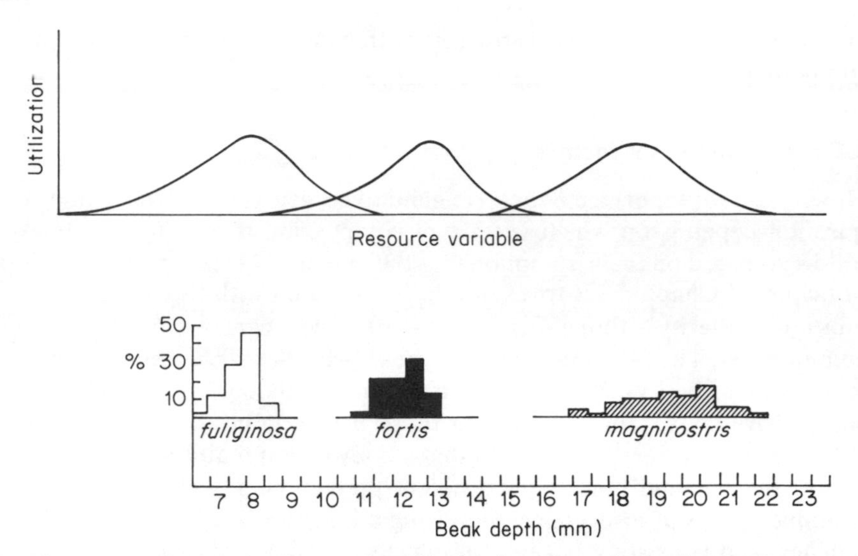

Figure 8.1 Regularly dispersed RUFs (top) and regularly dispersed morphological character distributions: the simplest possible view of a competitive guild

the spectrum so that the spectrum fills up with RUFs in some sequential manner—or if this does not happen, then there is a mechanism which ensures some sort of 're-shuffling' where an initially adjacent coexisting pair overlap to a degree that is not the maximum permitted under the limiting similarity criterion.

In addition to the above, if the data under consideration are morphological rather than dietary, then we must further assume that:

8. There is a straightforward correspondence between distributions of the morphological character and RUFs.

Although I am the proposer of the mechanistic competitive exclusion principle, a supporter of the idea of a (variable) degree of limiting similarity, and a 'believer' that some guilds in natural communities are competitive, I cannot accept as realistic the above eight assumptions, with the exception of number 6 and, in some communities only, number 2; and I suspect that many other proponents of niche theory would take a similar view. Thus it does not make sense to consider this simplistic picture any further, nor to devote much time to assaults on it by various critics from the 'anti-competition lobby'. Rather, we will concentrate on enquiring what sort of competitive guild structure might be expected on the basis of more realistic assumptions.

If we relax all of the above eight assumptions, except that of limiting similarity, we get a very different picture of guild structure, as shown in Figure 8.2. This 'structure' is much more difficult to distinguish from a random collection of RUFs, and indeed there is really only a single competitive effect

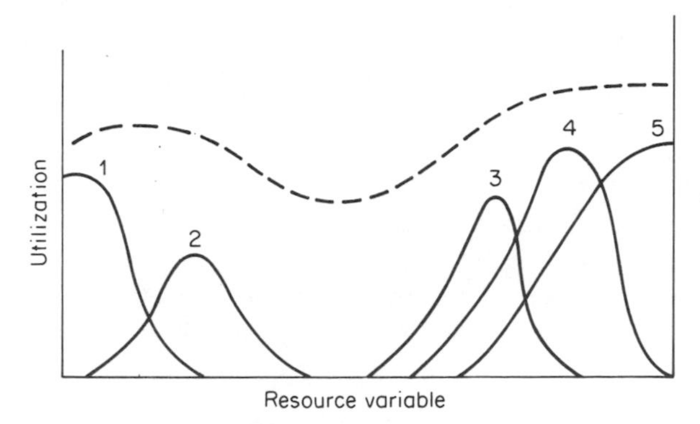

Figure 8.2 A 'realistic' competitive guild of five species. (Dashed line illustrates variation in resource abundance.)

on it, namely *no species are separated by less than the limiting similarity criterion*. Notice that this says nothing about how many will be separated by more than this criterion, nor about exactly what the criterion is (it could be $d/w = 1$ or some alternative value).

Taking this view, then, apparent examples of regular niche spacing, including the ground finches mentioned above and other case studies such as those discussed by May (1974b, Chapter 6), are seen as unusually simple guilds, not 'typical competitive guilds'. If the typical competitive guild is indeed as shown in Figure 8.2, then there will be difficulties in distinguishing the competitive effect on structure from others, and difficulty in testing the competitive effect because it is not clear exactly what limiting similarity value should apply. This is a depressing prospect, but it would seem dishonest to pretend that the situation were otherwise.

8.2.3 Among-guild patterns

If we compare competitive guilds of the same sorts of consumer organism utilizing the same kind of resource in different geographical locations, what sort of pattern is likely to emerge? One obvious answer is that in places where there is a longer resource spectrum (i.e. more resource types available), there should in general be a greater species diversity in the consumer guild. I use the term 'species diversity' broadly; it does not matter for current purposes whether we are considering simply the number of species ('species richness') or a diversity index which compounds richness with evenness of abundance.

MacArthur (1972) has formalized this relationship in the following way (see Figure 8.3). The length of the resource spectrum is R, the range of an RUF is denoted U, and U is divided into H and O which relate to (but are not exactly equal to—see Figure 8.3) the 'exclusive' and 'shared' parts of the

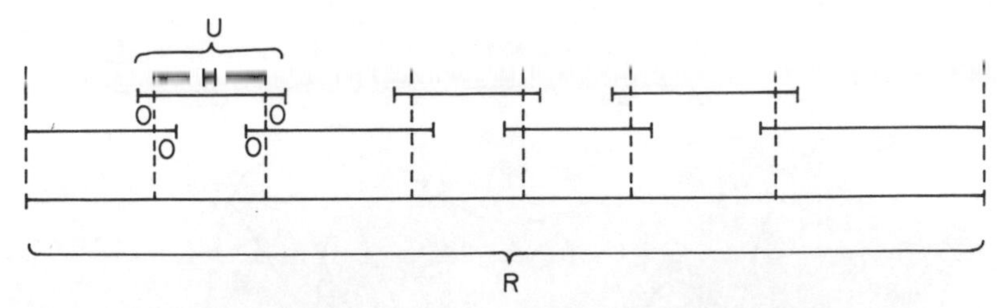

Figure 8.3 Utilization of resource spectrum of length R by seven consumer species. From MacArthur (1972)

RUF. This representation omits any information on the actual pattern of utilization by a particular species.

MacArthur reasons that in comparisons of guilds of the same kind of organism in neighbouring localities which differ considerably in R, differences in species diversity from place to place should largely be explicable in terms of the different R-values, as the values of U, O and H should not vary much. He then discusses a classic example where this is indeed the case, namely the relationship between bird species diversity (BSD) and foliage height diversity (FHD) revealed in an earlier study (MacArthur and MacArthur, 1961). This relationship, in which both diversities are measured by the formula $D = 1/\Sigma_i p_i^2$, is shown in Figure 8.4 (p_i is the proportion of individuals in the ith species).

The strong positive association between these two variables is clear, and its existence can hardly be disputed. (Even the relatively weak Spearman rank correlation test gives $r_s = 0.94, p < 0.01$.) Its interpretation, however, is much more problematical, and in a sense the problem is the opposite of the one we

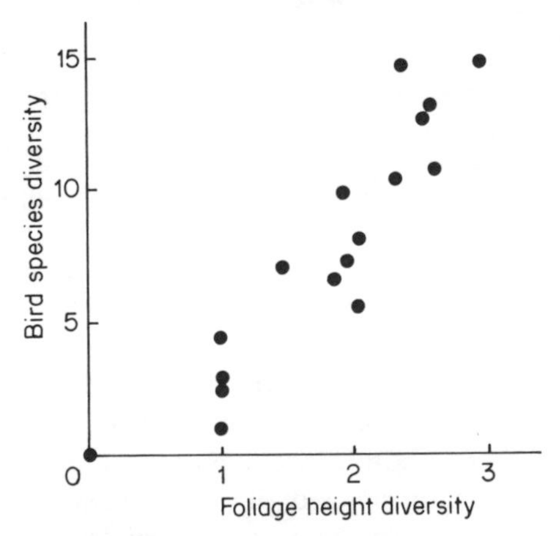

Figure 8.4 Positive association between bird species diversity and foliage height diversity. From MacArthur (1972)

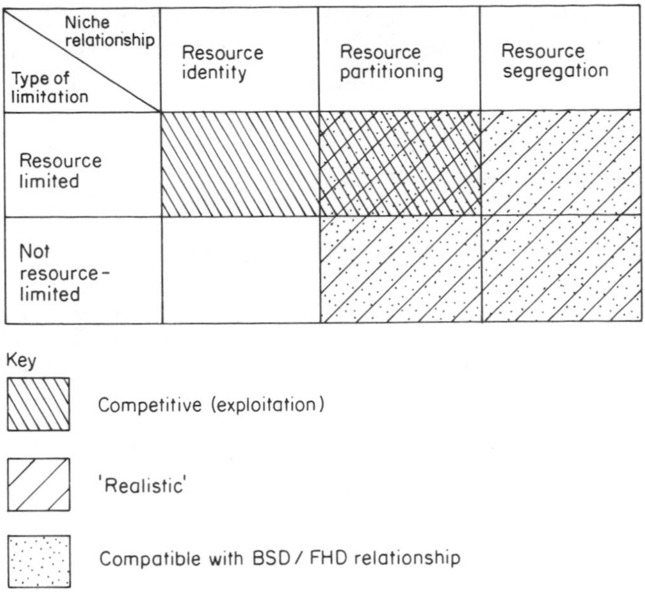

Niche relationship / Type of limitation	Resource identity	Resource partitioning	Resource segregation
Resource limited			
Not resource-limited			

Key

Competitive (exploitation)

'Realistic'

Compatible with BSD / FHD relationship

Figure 8.5 Illustration of the non-correspondence between the BSD/FHD association and competition

encountered when looking at within-guild structure. That is, it is not that this pattern is implausible, but rather that it is so plausible that it is predicted on the basis of most reasonable views of guild structure, both competitive and otherwise.

If we consider RUFs to be unidimensional (because it simplifies the argument) then there are three possible relationships between adjacent RUFs—identity, partitioning and segregation. Combining these with the dichotomy of resource-limited and non-resource-limited systems gives six possible situations, as shown in Figure 8.5. There are three relevant ways of dividing up these situations, all of which are indicated in the figure: competitive versus non-competitive; 'realistic' versus 'unrealistic'; and situations which are compatible with the BSD/FHD relationship versus those which are not. Notice that all four 'realistic' scenarios are compatible with the relationship but only one of these involves interspecific competition.

8.2.4 Summary

There is increasing evidence that at least some guilds are competitive, that is, that they are characterized by exploitative competition among some or all of their constituent species. However, with regard to patterns of structure within and among guilds, we have seen that some patterns that have been proposed for competitive guilds are rather unrealistic and would hardly be expected to be common in nature, while others are realistic but would be expected in all guilds—both competitive and otherwise. So far, we have seen no structural criterion by which competitive guilds can be distinguished, apart from the existence of competition itself and a limiting-similarity criterion whose precise nature is unknown. It is against this background that we examine some of the recent assaults on the concept of the competitive guild.

8.3 OTHER GUILDS

The concept of limiting similarity may be inapplicable to a guild for two distinct reasons, as noted earlier. It may be inapplicable because, although the guild is competitive, the competition taking place is not adequately described by niche theory. I will not discuss this possibility further, except to note that one recent subject of discussion here concerns the ability of niche theory to deal with situations involving highly asymmetrical competition (see Wilson, 1975; Lawton and Hassell, 1981, 1984; Persson, 1985; and also Chapter 10). I will concentrate instead on the alternative possibility: that limiting similarity is inapplicable because guilds are not competitive, and that apparent structure within and among guilds either is indistinguishable from the 'structure' that would be presented by a random collection of species or is caused by processes other than competition, for example predation.

8.3.1 Null guilds and null models

Over the past decade a substantial literature has accumulated on the use of 'null models' to analyse guild or community data. Two of the seminal articles precipitating the deluge of studies of this kind were those by Strong *et al.* (1979) and Connor and Simberloff (1979). More recent landmarks in the continuing debate over the usefulness of null models are the 'round table' issue of *American Naturalist* (1983; **122**, no. 5), the review article by Harvey *et al.* (1983), and the collection of papers edited by Strong *et al.* (1984). An early example in this field of what would now be called a null model is provided by Williams (1964, Chapter 9).

I will examine the approach of the null modellers by looking in some detail at the paper by Strong *et al.* (1979). These authors attempted to assess the evidence for what they call 'community-wide character displacement', and it will help if we realize at the outset that this is a complete misuse of the term character displacement—what they were really examining was the evidence for regular spacing of niches, as reflected in distributions of morphological characters. There is no necessary connection with coevolutionary processes occurring in sympatry, because regular spacing, if it exists, can be produced simply by differential competitive exclusion of those immigrants that are too close to an already resident species (or, for that matter, differential exclusion of the residents).

Given a series of morphological distributions, one per species, strung along an axis of 'character value', there are two possible extremes of interpretation. One is to note that there is some degree of spacing between the means of adjacent species, to describe this spacing in terms of the larger mean divided by the smaller, and to interpret the resultant ratio (1.2, 1.3, 1.4 or whatever) as being related to limiting similarity—an approach that can be attributed to Hutchinson (1959). The alternative extreme is to note that there is some variation in the ratios when different species pairs are considered, and to reject competition on this basis because variable spacing is inconsistent with the simplistic version of the competitive guild given in the previous section where $d_1 = d_2 = d_3$ and so on.

While I am unaware of any author actually reasoning in this latter way, it is the logical opposite of the 'Hutchinsonian ratio' approach. One view says that character ratios are evidence for competitive constraints because they are not very different from each other; the alternative view holds that the ratios are evidence for the absence of competitive constraints because they are too different from each other. In neither case is the reasoning satisfactory because there is nothing sensible with which to compare the data—in a sense there is no null hypothesis. We can ask the question 'what sort of distribution of character ratios would there be in a non-competitive guild? and, so far in our examination of this problem, there is no sign of an answer.

The advance made by the null modellers—and there is no doubt that it was an advance, despite the problems to be discussed below—was that they attempted to provide an answer to the above question, that is, to provide something with which the actual data could be compared. Thus, they argued, it was possible to test, based on the magnitude of the departure between observed and 'expected', whether there actually was a competitive effect.

The process can be exemplified by the treatment by Strong *et al*. (1979) of the bird species of the Tres Marias Islands, Mexico. Within each family, they arranged species in order of size (independently with respect to two characters—wing length and culmen length) and calculated character-ratios between 'adjacent' species. So, given S species of a particular family on an island, there were $S - 1$ ratios. In theory, each such ratio can be anything in excess of 1; in practice they all fell in the range 1 to 3. Observed data were calculated from sympatric species on individual islands. For comparison with this data, the authors randomly selected the appropriate number of species from a 'source pool' represented by bird species present on the adjacent area of mainland. They calculated ratios for each random draw; their expecteds were obtained by averaging 100 such draws in each case. Comparing observed with expected for wing length gave 14 out of 30 observed ratios greater than expected and for culmen length 18 out of 30. As Strong *et al*. (1979) state, 'these proportions are not different from random expectations'.

As I have stressed, the advance of such an approach over many earlier studies lies in its refusal to assume competitive structuring of the community, and its insistence that, instead, we should test for such structuring by comparing the actual data with what has become known as a null model. However, we need to enquire whether the form of testing used is (a) valid and (b) sufficiently powerful.

Several criticisms of the null model approach can be, and have been, made. First, what is assembled for comparison are not null guilds but null 'taxonomic assemblages' in the sense of Jacsic (1981) or null 'bird communities' to use the commoner, and less correct, phrase. These are much too broad groupings for competitive effects to be dominant. As Colwell and Winkler (1984) put it: 'If interspecific competition occurs in a community it presumably will be strongest and hence most detectable in a small subset of ecologically similar species. Including other less similar species in an analysis runs the risk of obscuring the effects of competition'.

Related to this is the criticism—also made by Colwell and Winkler (1984)—that the tests used to compare observed and expected values are too weak. Given that what is really being tested is whether morphology is relevant *at all* to species co-occurrence (either through competition or through other means) and that the result of the tests of Strong *et al*. (1979) and other null modellers are negative, this criticism is clearly a serious one. Can we really believe that morphology is that irrelevant? I, for one, cannot.

Another objection to the work of Strong *et al*. (1979) is that using morphological measurements to infer RUFs is highly suspect. While morphology is not ecologically irrelevant, the eco-morphological relationship is not likely to be a simple one. One problem that is readily apparent, yet rarely explicitly admitted, is that morphological character distributions, at least when based on the adult subset of a population, have WPC = 0, BPC = 1. That is, each individual is a 'point' on the axis of character value, not a range. This contrasts very strongly with RUFs, where the situation to be expected is for both WPC and BPC to be fractional; and indeed it often seems likely, given extensive flexibility of the feeding behaviour of individuals, that WPC > BPC.

Finally, it must be stressed that a null model approach such as that of Strong *et al*. (1979) is only readily applicable to a study of islands and neighbouring mainland. Given this situation, the observed data come from the islands, the 'expecteds' from the mainland. Studies of island archipelagoes on their own, with expecteds being generated by a re-shuffling of the observed data on the archipelago, opens up serious problems additional to those discussed above. These problems have been described by Gilpin and Diamond (1984) in relation to Connor and Simberloff's (1979) 'null study' of an archipelago system. Continental studies where we have mainland but no islands (except habitat islands) cannot even be approached by the usual null model method.

Although it would be possible to continue here listing more criticisms that have been levelled at the null model approach, it hardly seems useful to do so. Rather, I will end this section by (a) noting agreement with Gilpin and Diamond's (1984) concern that null models should not be equated with the statistical concept of a null hypothesis, which is of a very different nature and

(b) re-emphasizing the main strength and weakness of the null model approach. Their main strength is that they have caused a more critical view of patterns of community structure, and have helped to reduce the tendency of some authors to attribute almost any apparent aspect of structure to competition, without enquiring if the structure is real or competition occurring. The main weakness of null models is that in many cases, including Strong *et al.* (1979), the result of comparing the model with reality was negative—that is, natural communities were not distinguishable from random assemblages of species. Whether or not we accept that competition is important as an agent of community structure, we cannot deny that other real biological processes are important, including migration, predation and the general 'adaptedness' of a species to a particular island or habitat. Failure to distinguish natural communities from randomly-assembled ones can, therefore, only be interpreted as weakness in the discriminating power of the method employed.

8.3.2 Predator-limited guilds

While the idea that guilds are entirely random assemblages of species is difficult to believe, the possibility that many guilds are not resource-limited, and consequently not competitive, despite considerable niche overlap, is plausible (and also testable). Indeed, the question of whether guilds are generally limited by resources or by predators/parasites can be regarded as one of the most fundamental ones in the search for an understanding of guild structure.

While many of the field studies in which competition has been proposed as an important structuring force were based on populations of vertebrates (particularly birds), proposals that guilds are predator limited have often been based on studies of insect populations—in particular, populations of phytophagous (= herbivorous) insects. The possibility that these two groups (birds and insects) differ in this respect was noted long ago by Lack (1968, p. 136) who makes reference to a comment by G. C. Varley that the competitive effects discerned by Lack's study of Darwin's finches may not apply to insect populations.

More recent claims that guilds are predator limited and hence not subject to exploitative competition have also tended to originate from studies of phytophagous insects (e.g. Lawton, 1984; Seifert, 1984; Strong, 1984). I will consider in some detail Lawton's (1984) study of bracken-eating insects, both because this is based on a fairly well-defined, 'isolated' guild, and because Lawton uses several different arguments all of which seem to point to lack of competition and lack of competitively-induced guild structure.

Lawton (1984) describes an extensive set of data resulting from studies of bracken/insect guilds in the British Isles, Arizona, Hawaii and Papua New Guinea. Since bracken is currently thought to be a single species throughout its geographical range (*Pteridium aquilinum*), the resource part of the guild is remarkably simple and consistent. The number of insect species utilizing it varied from none, in Hawaii, to 39 in Britain. In general, the species come from several insect orders, so they are widely dispersed taxonomically, but within this wide taxonomic spread there is considerable 'clumping' as the species-lists include many pairs and groups of congeners.

Lawton provides four lines of evidence suggesting that these guilds are not resource-limited and do not exhibit exploitative competition among their constitutent species. Two of these lines of evidence focus on a search for overall 'density compensation', (this term refers to the tendency of a population to increase in number when other competing species are absent), and two on comparisons of distributions and densities of closely-related species pairs. I will deal with these two aspects of the study in turn.

The weakest piece of evidence concerns density compensation in a series of guilds from different sites in the North Yorkshire Moors. This group of sites (like almost any such group, for any taxon) shows a positive relationship between number of species and area. In the smaller sites, with fewer species, the remaining species do not show the increased densities that might be expected if interspecific competition is occurring. However, as Lawton points out, the bracken fronds tend to be smaller in the smaller sites, which makes this sort of comparison difficult to interpret.

Comparison of an intensively-studied British site (Skipwith, Yorkshire) with the Sierra Blanca site in Arizona is more satisfactory in this respect, because the site of lower diversity, Sierra Blanca, has slightly larger fronds. The finding that there was no density compensation in this comparison cannot therefore be explained in terms of the smaller number of species at Sierra Blanca having less food. While I have some qualms about comparisons of two such distant sites, between which so many variables must differ, it has to be admitted that the available evidence does not point to food limitation.

The work on overall density compensation can be criticized from the same viewpoint from which the study by Strong *et al*. (1979) was criticized—namely that it may dilute competitive effects by considering a wide taxonomic assemblage. However, Lawton (1984) also looked for evidence for competition in pairs of closely related species and it is this work to which we now turn. First, Lawton looked for evidence of the 'checkerboard' pattern of distribution, which is widely regarded as indicative of competition between two species. However, in a large number of pairwise comparisons, presence of one species either did not affect the likelihood of presence of the other, or in some cases increased it—the latter result being similar to that of Williams (1961, Chapter 9).

Where two species are found together at a local level in several places, their densities should, if they are in competition and other factors are constant, be negatively related. Lawton (1984) found no evidence for this; abundances per frond were in almost all cases not significantly associated. Also, while some of the comparisons involved different genera, many of them involved a pair of congeners.

In addition to looking for evidence of resource limitation/competition,

Lawton examined the pattern of niche relationships within and among guilds. In particular, he looked for evidence of convergence in the overall pattern of resource use from place to place. Apart from a general tendency at all sites for the pinnae ('leaves') to be attacked most, there was little evidence of convergence, and much evidence of empty niche space. This pattern certainly provides no evidence for competition but, bearing in mind the 'realistic competitive guild' discussed in the previous section, it does not provide very strong evidence against competition either.

In summary of Lawton's (1984) extensive work on this system (see also Lawton, 1976), it can be said that while no one argument is conclusive on its own, and while it would be desirable to have some experimental studies to back up the observational ones, the various lines of available evidence, taken together, do suggest that this guild is not a competitive one. Moreover, the studies of Strong (1984), Seifert, (1984), Lawton and Hassell (1984) and others suggest that the bracken guild is by no means unique in this respect.

8.4 SOME OF EACH

While the nature of the structure to be expected of both competitive and noncompetitive guilds remains to be determined in more detail, it is becoming clear that some guilds in nature are indeed limited by their resources, while others appear not to be. Can we currently make any bolder statement than this?

A rather extreme approach is that of Pimm (1980, 1984) who asserts that food limitation ('donor control') is rare in nature and, by implication, that instances of competitive guilds are rare exceptions to a general rule of limitation by predators or parasites. Lawton (1984), though adopting a less extreme stance, argues that the non-competing insect herbivores of bracken are typical of phytophagous insects in general, and perhaps even of a wider group than that. He concludes: 'a clear majority of the world's biota must therefore behave this way'.

Rather than attempting to generalize overall about which form of limitation is more common, it may be more fruitful to attempt to generalize about 'where' we will tend to find one or other sort of guild. One possibility to which I have already alluded is that certain taxa may more often exhibit exploitative competition than others (e.g. birds, compared with insects). However, a more potentially interesting kind of generalization (which may interact with the taxonomic one) is one phrased in terms of trophic levels.

The first attempt at such a generalization was that of Hairston *et al*. (1960; see also Slobodkin *et al*., 1967; Hairston, 1985). These authors proposed that, in general, plants, carnivores and detritovores would be food-limited, while herbivores would tend to be limited 'from above'. More specifically, they argued that the actual trophic levels, each considered as a unit, were limited in the way just described, and that individual populations within each level would tend to be but need not be limited in the way that their level was limited. This is important, because, as I stressed at the beginning of this chapter, the trophic level is too large an ecological unit in which to look for any absolute generalization regarding the 'direction' of population limitation. Rather, any differences between trophic levels are statistical in nature.

An alternative to the proposal by Hairston *et al*. (1960) is suggested by Pimm (1980). In keeping with his assertion that resource limitation is rare overall, Pimm claims that the only large trophic unit to which it is likely to apply is the detritovore/decomposer system. Of course, at this stage, both of these are simply competing assertions, either or neither of which may turn out to be true, and both of which are compatible with the growing feeling that herbivore populations are generally not food limited. I have to admit a leaning towards the generalization by Hairston *et al*. (1960) rather than Pimm's; but this is based on little more than an intuitive feeling—and we all known how reliable those are!

Two final points need to be briefly mentioned. First, predators can be important agents of coevolution and community structure regardless of whether they limit populations, while exploitative competition does not occur (and hence is not an agent of structure) if resources are not limiting (Jeffries and Lawton, 1984). Second, the whole idea of a single limiting factor, with populations being limited either by their resources or by their consumers but never by both, may be an oversimplification, at least in some cases (Taylor, 1984). I will return to this point in Chapter 10.

Chapter 9

The Niche in Long-term Evolution

9.1 INTRODUCTION

This is the most autonomous chapter of the book, and it may be profitable to start it by enquiring why. There are two main reasons. First, although they have many points of contact, ecological theory and evolutionary theory are largely separate endeavours with, for the most part, different approaches and different goals. Even evolutionary ecology, despite its title, does not truly bridge the gap between the two. Since this is an evolutionary chapter embedded in a largely ecological book, it is necessarily 'out on a limb' to some extent.

While the book has a predominantly ecological flavour, two previous chapters (3 and 6) briefly outlined the concept of polymorphism. The second reason why the present chapter is decidedly autonomous is that the wider evolutionary issues which are discussed herein are not only largely detached from ecological theory, but also largely detached from the narrower evolutionary issues addressed in population genetical theories on the dynamics of polymorphism.

Whether the current split between the theory of polymorphism and broader-scale evolutionary theory is temporary and undesirable or permanent and, in a sense, desirable, hinges on the following question. Are major evolutionary steps, such as the origin of a new species, compound events decomposable into a collection of smaller, constituent events, such as transient polymorphisms? The conventional answer, championed by architects of the 'modern synthesis' (or neo-Darwinism), such as Mayr (1963), is a straightforward 'yes'. The alternative possibility is that speciation and other major evolutionary steps are fundamentally different from the microevolutionary processes going on within existing species. This view is put increasingly often by a diverse collection of evolutionary biologists who have rebelled against what they see as the neo-Darwinian straightjacket. For example, P. G. Williamson (1981) states that 'speciation is a qualitatively different phenomenon from gradual, intraspecific microevolutionary change'.

While I have some sympathies with (and some objections to) both of the above views, a book on the ecological niche is hardly the place to try to resolve this major conflict within evolutionary theory; and at any rate, I have already examined this conflict elsewhere (Arthur, 1984). In the present chapter, I will adopt a much more limited approach, and will concentrate on the role of the niche in the origin of species and superspecific taxa. Even taking this approach does not completely avoid controversy, because one possible application of niche theory to the origin of a new taxon involves 'saltational' change, which is denied any importance in evolution by strict neo-Darwinists. However, I will postpone discussion of this possible application until Section 9.4, and will concentrate, in the intervening sections (9.2 and 9.3), on making some more fundamental, and less controversial, statements about the role of the niche in major evolutionary events.

Two final comments are necessary before closing this introductory section. First, it is clear that many evolutionary biologists using the term 'niche' are taking a fairly broad interpretation of the term, often akin to Elton's (1927) usage where a niche represents the role of a population in a community. Other evolutionary biologists use 'niche' primarily to describe resource utilization, as I have done herein. The extent to which niche considerations are useful in evolutionary theory depends, of course, on which niche concept is employed. If we take a broad view of what constitutes a niche, then the concept is very widely applicable but in a sense trivial, since every major evolutionary change involves a niche shift of some sort. If we stick to the narrower, resource-based niche concept, then evolutionary changes describable as niche shifts are less common in their occurrence but are in a sense more meaningful when they do occur.

My final 'introductory' comment is that inasmuch as it connects with the theory of interspecific competition, this chapter tells us that major evolutionary changes are associated with a lack of such competition. This conclusion seems to hold regardless of whether the changes involved take place gradually, as in the conventional neo-Darwinian view, or saltationally. Compared to the kinds of evolutionary shift possible in the absence of interspecific competition, those coevolutionary changes, such as character displacement, which may occur in its presence, appear merely as a kind of evolutionary 'fine tuning'.

9.2 THE NICHE AND ALLOPATRIC SPECIATION

Although our knowledge of speciation is very limited, especially when contrasted with our understanding of intraspecific evolutionary processes, there is a general consensus, at least among evolutionary zoologists, that speciation most commonly occurs between allopatric populations. Lewontin (1974) puts it as follows: 'If there is any element of the theory of speciation that is likely to be generally true, it is that geographical isolation and the severe restriction of genetic exchange between populations is the first, necessary step in speciation'. Allopatric speciation is also thought to be common in plants, though botanists see it as one of a wider range of speciation mechanisms, including speciation by hybridization of already-existing species (see Grant, 1981).

As Lewontin's comment makes clear, optimal conditions for speciation include a geographical barrier to gene flow as well as the mere fact of allopatry. Depending on where such a barrier falls in relation to the distributional range of the parent-species, a continuum of possible geographical patterns of speciation may be envisaged. At one extreme is the approximately equal split that would be involved if a holarctic species diverged into nearctic and palaearctic daughter-species as a result of different selective regimes on different sides of the Atlantic. At the other extreme is the case of speciation via a 'peripheral isolate'—that is, a small population inhabiting a restricted area (perhaps an oceanic island) separated by a barrier of some sort from the rest of the range of the parent-species.

Speciation involving oceanic islands has been particularly intensively studied. Certainly, the best examples of recent allopatric speciation in animals derive from studies of island systems. Many of these examples are discussed by M. Williamson (1981). Perhaps the best known are the cases of Darwin's finches in the Galapagos and the Hawaiian Drosophilidae. Both of these case studies have been well described elsewhere (see Lack, 1947 and Grant, 1986, for the Galapagos finches; Carson *et al.*, 1970, and Williamson, 1981, Chapter 8, for Hawaiian *Drosophila*), and consequently I will not discuss them at length. However, I will give a brief outline of speciation in the Galapagos finches, with particular reference to their feeding niches. This will help to illustrate the role of the niche in the speciation process in general.

Ironically, although Darwin's finches have told us much both about the importance of the dietary niche, and about the allopatric form of speciation, the aspects of this case study that are most informative about allopatric speciation tend to be relatively uninformative about feeding niches, and vice versa. This is because the details of speciation are most clearly seen where it has happened recently—but recently-separated species will show relatively slight differences in diet (and in other characteristics). On the other hand, the array of niches into which Darwin's finches have radiated is most clearly seen by contrasting the major branches of their evolutionary tree, where diets are quite distinct. I will adopt this second, broader, strategy first.

'Darwin's finches', alias the subfamily Geospizinae, are endemic to a group of islands off the western coast of South America, comprising the Galapagos archipelago and the solitary Cocos Island (see Figure 9.1). There are five genera —*Geospiza*, *Camarhynchus*, *Platyspiza*, *Certhidea* and *Pinaroloxias*. The last of these is represented by a single species, *P. inornata*, found only on Cocos. Its distribution is thus mutually exclusive to that of the other four genera, which are found only on the Galapagos.

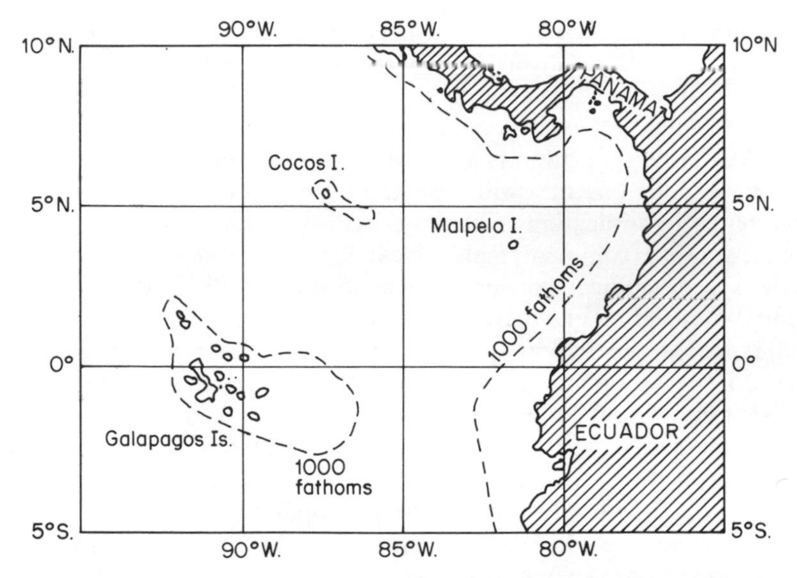

Figure 9.1 Location of the Galapagos archipelago and Cocos Island. Reproduced by permission from Lack (1947) *Darwin's finches* published by Cambridge University Press

Although the evolutionary tree of the Galapagos genera is not known with certainty (as is the case with the vast majority of such trees), there would appear to be five main branches—two for *Geospiza* and one each for the other three genera—all radiating out from an ancestral species not unlike *Geospiza difficilis*. These branches are as follows (based on Lack, 1947, 1971, with some modification):

1. The ground-finch branch. This includes the large, medium and small ground finches—respectively *G. magnirostris*, *G. fortis* and *G. fuliginosa*. These three species have heavy, finch-like bills and feed predominantly on the ground on a variety of seeds. As discussed in Chapter 7, there is a possible case of character displacement involving *G. fortis* and *G. fuliginosa*.
2. The cactus-feeding branch. This branch comprises the two remaining *Geospiza* species, *G. scandens* and *G. conirostris*. Although not quite separate in diet from the ground finches, both of these two species consume cactus food. The diet of *G. conirostris* appears not to be well known, perhaps because this species is not found on the central group of islands. *G. scandens*, however, is known to consume flowers and fruits of the prickly pear *Opuntia*.
3. The *Certhidea* branch. The single species of this genus, *C. olivacea*, has, in Lack's (1971) words, 'a rather slender, straight, tweezer-like beak, like

that of a warbler, for picking small insects off leaves and fine twigs, which the other Geospizinae rarely if ever do'.

4. The main insectivorous branch. This includes the five species of the genus *Camarhynchus* (sometimes divided into a subgenus of that name and a second subgenus called *Cactospiza*). These birds hunt for insects in a variety of places including under bark, in fissures (using a small stick as a tool) and in rotting wood.
5. The *Platyspiza* branch. This is another monospecific genus. The species concerned, *P. crassirostris*, eats a variety of 'vegetarian' foods, including buds, leaves and fruits.

As mentioned above, the present-day Galapagos species that seems to have remained closest to the ancestral form is *G. difficilis*. But where did the ancestral species come from? The answer to this question is not entirely clear, though there seem to be two main possibilities. An unknown and now extinct finch species (family Fringillidae) may have migrated to the Galapagos archipelago from the South American mainland (Lack, 1947). If so, then Darwin's finches really are finches, and the species on Cocos, *P. inornata*, was derived later from some branch of the adaptive radiation described above—perhaps the *Certhidea* branch. The alternative possibility (Lack, 1968; Harris, 1973a; Williamson, 1981) is that Darwin's finches arrived on the Galapagos via Cocos Island, and that *P. inornata* (or at least an early version of it) was ancestral. Harris (1973a) suggests that *P. inornata* is in turn a descendant of the South American mainland species *Coereba flaveola* (family Parulidae). If so, then, ironically, Darwins finches are not finches at all, but rather an aberrant group of wood warblers!

However the ancestral finch arrived in the Galapagos archipelago, and whatever its evolutionary origin, we can imagine the situation it encountered. It was confronted by a group of small volcanic islands, all different to some extent in size, physical characteristics, distance from other islands, and biota. There were presumably few other species of land birds on the archipelago. [Even now, there are only eighteen such species (Harris, 1973b), so the finches, with thirteen species, make up nearly half of the avifauna.] There was thus a considerable amount of empty niche space, the nature of which differed from island to island. We can picture occasional migrations between islands establishing new populations, all originally of the same species, but all being subsequently selectively modified in a direction determined by local island conditions, particularly the available food supply. The process so far described would lead, at most, to one (different) species per island. Since most islands now have several species, we must imagine that re-migrations, occurring after sufficient genetic divergence to prevent interbreeding, produced the more complex finch faunas that we see today; perhaps with some 'coevolutionary fine-tuning' of some pairs of species in neo-sympatry.

The account given in the previous paragraph is, of course, entirely speculative, and the overall pattern of migration, divergence, speciation and re-migration will presumably never be known. However, if we take a more restricted view, the pattern of speciation can sometimes be seen more clearly.

This is particularly true if we focus on the pair of species *Camarhynchus psittacula* and *C. pauper*, which have diverged from each other relatively recently. I will refer to this pair of species, including three subspecies of the former, as the *C. psittacula* species complex. This complex inhabits the central Galapagos islands, but within this overall area different members of the complex are found on different island groups (see Figure 9.2). The southern 'group' consists of only a single island, Charles, and is inhabited by *C. pauper* and *C. psittacula psittacula*, which do not interbreed. The other three groups of islands are inhabited, in the sequence W, N, E, by *C. psittacula affinis*, *C. psittacula habeli* and *C. psittacula psittacula*. It seems almost certain that *C. pauper* was originally an island form or subspecies of *C. psittacula*; but it has clearly now diverged to the extent that it is a separate species in its own right and so has remained distinct despite the secondary migration of *C. psittacula psittacula* on to Charles. This is as clear a case of recent allopatric speciation as is available from any taxonomic group.

It seems likely that speciation in other groups of animals and in other places is not fundamentally different to the example given above. Admittedly, Darwin's finches give us an unusually clear picture of the speciation process, partly because of the discrete nature of the island system they inhabit.

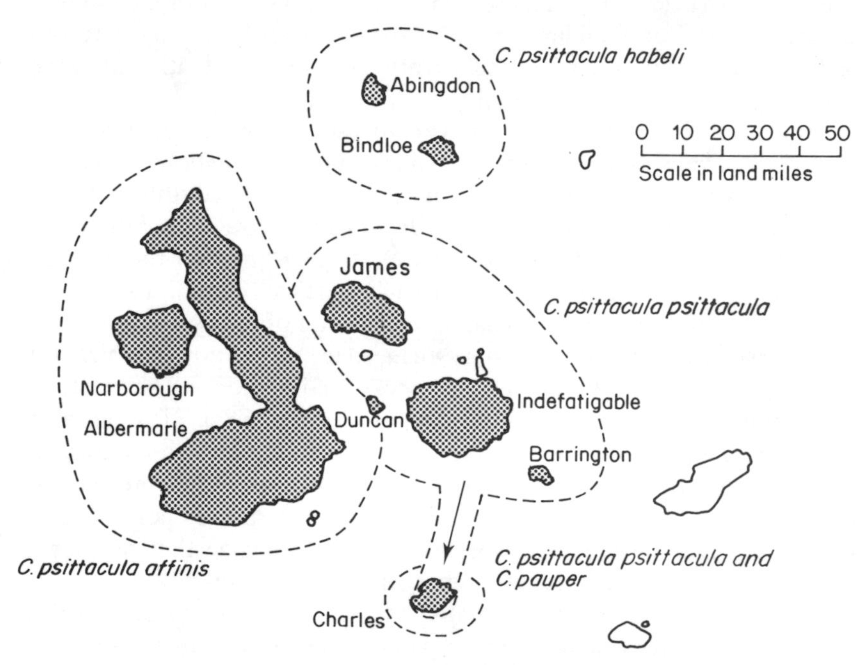

Figure 9.2 Distribution of the *Camarhynchus psittacula* complex on the central Galapagos islands. Reproduced by permission from Lack (1947), *Darwin's Finches*, published by Cambridge University Press

However, the atypicality is in our ability to visualize the process, not in the process itself.

Of course, as with any general process illustrated by a single example, there is a danger that we may overemphasize certain peculiarities of the example concerned. In this context, we should be careful not to assume that modification of the food niche is a necessary part of speciation. There is no reason why, in theory, the speciating entities should not remain the same in this respect. However, in practice, if allopatric speciation is the rule rather than the exception, then some divergence in food niches seems likely since different localities will usually differ qualitatively and/or quantitatively in the range of foods available. Also, it must be stressed that the relative availability of different foods will mould the food niche evolutionarily *regardless of whether or not food is limiting*. Exploitative competition will only mould niches where resources are indeed limiting (since otherwise it does not occur), as pointed out by Jeffries and Lawton (1984) and briefly discussed at the end of the previous chapter; but the influence of food supply is much more pervasive than the influence of competition.

Indeed, one of the main messages of the finch story is that lack of interspecific competition is important for major evolutionary shifts such as have occurred in the overall adaptive radiation of the Geospizinae. If the ancestral finch had arrived to find the Galapagos archipelago populated by a great diversity of passerine land birds of other families, it might well have been an evolutionary dead-end. If this rather negative message for competition is true, it carries with it a more positive one. If new evolutionary directions are initiated largely when there is a paucity of fairly closely related sympatric species, it suggests that when such species are present, they do in fact compete. If all species populations were permanently sub-equilibrial in relation to their food supply, why should independent but geographically coincident adaptive radiations not occur in ecologically similar groups? Why, for example, has no other group of birds radiated out, in the Galapagos, to provide us with forms consuming a similar range of foods to the Geospizinae?

Coupled with the idea that lack of interspecific competition tends to be associated with evolutionary shifts goes the idea that occurrence of competition is associated with evolutionary stasis. That is, in a species-rich system, the sort of selection stemming from interspecific competition may most often be stabilizing. This hypothesis (and that is all that it is) seems to run contrary to the idea of competition being an agent of directional selection, as encountered in the process of character displacement. However, the two ideas are not mutually exclusive. One satisfying picture which combines the two is that of a gradual increase in the species-richness of a group as its adaptive radiation proceeds, involving sequentially:

1. Major evolutionary shifts in the absence of competition.
2. Coevolutionary fine tuning as pairs of species come into competitive contact.

3. Stabilizing selection predominating as each species becomes sandwiched either between two others or between one other and the end of a resource spectrum.

Finally, let us return briefly to the issue of the different geographical patterns of allopatric speciation which were mentioned at the beginning of this section. In particular, let us consider two such patterns—archipelago speciation, as in the Geospizinae, and speciation of a 'peripheral isolate'. This latter term is usually intended to describe a situation where a small, far-flung population from a species whose range is otherwise extensive and coherent, splits off reproductively from its parent-species. For example, if the outlying and partially-speciated population of *Drosophila pseudoobscura* at Bogota, Colombia, (Prakash, 1972), completes its split from the rest of that species (whose distribution is confined to western North America), this would provide a good example of 'peripheral isolate' speciation.

Although there are several differences between archipelago speciation and speciation by peripheral isolate, one thing that both have in common is the involvement of at least one population with the following features:

1. A large distance or geographical barrier separating it from all other conspecific populations.
2. Occupation of a restricted geographical area.
3. High likelihood of having been established from just a few 'founders'.

All three of the above attributes have important consequences for the speciation process. The first will tend to mean that the selective regime in the isolated population is atypical for that species and that gene flow will be low or non-existent. The second will lead to a reasonably homogeneous selective regime, thus helping to prevent different selective pressures diluting each other's effects, as may happen in a large, ecologically heterogeneous region. The third brings in the possibility of Mayr's (1942, 1954, 1963) 'founder effect', in which selection for external reasons may be aided both by the random genetic changes associated with small samples and by selection occurring for 'internal' reasons, induced by these initially random changes. In short, archipelagos and peripheral isolates present a variety of stimuli towards speciation, and we should perhaps expect that these two patterns of speciation would thus tend to predominate. This has implications for the fossil record, especially if the species so formed later becomes more extensive in its range. Basically, sequences of fossils taken from most parts of the enlarged range of the new species will show merely its sudden appearance by migration—something I have called an 'ecological punctuation' (Arthur, 1984). They will not show the rate at which it evolved from its parent species; and the question of how rapid that rate was then remains unanswered unless the geographical site of speciation can be established. If it can be established, and if the rate was rapid, then we have an 'evolutionary punctuation' at that locality. The theory of punctuated equilibrium (Eldredge and Gould, 1972;

Gould and Eldredge, 1977) is actually a theory about *ecological* punctuations (and about stasis between punctuations), as made clear by Gould (1977, p. 21)—though it is often interpreted otherwise.

9.3 ECOLOGICAL EQUIVALENTS

The concept of a pair or group of species that are ecological equivalents pervades the literature of both ecology (e.g. Odum, 1971; Emlen, 1973; Krebs, 1985) and biogeography (e.g. Cox and Moore, 1985). Odum (1971) gives the following as his 'statement' on this topic: 'Organisms that occupy the same or similar ecological niches in different geographical regions are known as *ecological equivalents*'. Emlen (1973), while also using this term, introduces another, synonymous one, namely *ecospecies*. I will avoid use of the latter partly because it is an unnecessary addition to the literature (who needs synonyms, in a field already overburdened with terminology?), and partly because there are too many kinds of 'species' already anyway (biospecies, morphospecies, palaeospecies, etc).

While there is no clear delineation of what can or cannot constitute a 'geographical region' for the purpose of recognizing ecological equivalents, an obvious choice of geographic unit is the biogeographical realm. These realms, often known as Wallace's Realms in honour of A. R. Wallace, are six in number, and have the following names: Nearctic, Neotropical, Ethiopian, Australasian, Palaearctic and Oriental. The first four correspond approximately (but by no means exactly) to the continents of North and South America, Africa and Australasia. The last two together correspond roughly to Eurasia, with the Oriental Realm being that part of Eurasia to the South-East of the Himalayas and associated mountain ranges. A very readable account of Wallace's Realms, together with their biogeographical significance, can be found in Elton (1958, Chapter 2).

Because of their largely separate evolutionary history over the past 50 million years or so, the six Realms have considerably different biotas (though human activities are reducing the differences—see Elton, 1958). Thus, given a particular broadly defined feeding niche, different species, and not very closely related ones at that, occupy that niche in the different Realms. An often-cited example is the 'niche' that is filled by large plain-dwelling herbivores. Before the advent of man, this niche was filled by bison in the Nearctic realm, kangaroos in Australasia, antelopes, gazelles and zebra in the Ethiopian Realm, and wild horses and asses in the Palaearctic. Ecological equivalents may also be recognized in more tightly defined niches, and examples from birds and lizards are given, respectively, by Karr and James (1975) and Pianka (1975). At the extreme level of detail represented by precise position within a guild, it may not always be possible to recognize ecological equivalents (see Lawton, 1984), but this does not detract from the usefulness of the concept at a grosser level.

What does the existence of ecological equivalents tell us? First, and most obviously, it tells us (yet again) that geographical barriers are of considerable importance in evolution. Given no barriers and free dispersal, there would presumably be no ecological equivalents at all. Second, it tells us that parallel or convergent evolution can fashion organisms to perform a similar task regardless of the phylogenetic starting point. Finally, it reinforces the idea that lack of competition is important for major evolutionary change. Given some empty niche space *and sufficient evolutionary time*, something will evolve to utilize it.

It is perhaps worth noting that the evolution of ecological equivalents in different Realms is in a sense the opposite of the adaptive radiation of a restricted group in a restricted area, such as the Galapagos finches. In the first case organisms with different evolutionary histories are evolving so as to have similar niches. In the second, organisms with a recent common ancestor are evolving apart and ending up with quite different niches. The fact that these two 'opposite' scenarios seem to tell similar stories—particularly with respect to the importance of barriers and a lack of competition—is reassuring.

9.4 THE CONCEPT OF *N*-SELECTION

For most of the history of evolutionary biology from Darwin onwards, a belief in gradual evolution has predominated. That is, at the phenotypic level, evolution has been envisaged as a slowly shifting mean value in the population, brought about by changes in gene frequency at loci which jointly produce the phenotypic character value but are individually 'invisible'. Also for most of the history of evolutionary biology the predominant, gradualistic view has been challenged by a diverse collection of dissenters.

The dissenters are often treated as a homogeneous group, but in fact their proposals differ from each other as well as from the conventional Darwinian view. These differences are most clearly seen in terms of the role each theory gives to macromutations in evolution. (These are mutations of large phenotypic effect, but not necessarily involving a large change in the genetic material.) Both de Vries (1905) and Goldschmidt (1940) argued that macromutations were typically involved in speciation, and that the production of new species was therefore quite distinct from intraspecific change (see also comment by Williamson, 1981, in Section 9.1). Others have argued that macromutations are *rare but important* evolutionary events, not involved in the typical speciation, but involved in the origin of some higher taxa at the order, class or phylum levels (Arthur, 1984). The 'liberal' neo-Darwinian view is that macromutations are *rare but unimportant* involving only a few 'special cases' such as the evolution of pigmentation patterns. Finally, the 'strict' neo-Darwinian view (which is patently false), is that macromutations never contribute to evolution or, as Darwin (1859) himself frequently put it, *natura non facit saltum*.

A notable absence from the preceding paragraph is the theory of punctuated equilibrium (Eldredge and Gould, 1972; Gould and Eldredge, 1977).

The reason for its omission in this context is that it is in a sense 'agnostic' about the role of macromutation, as pointed out by Gould (1982).

In the previous sections of this chapter I have tacitly assumed that all evolutionary changes, including speciation, take place in a conventional neo-Darwinian way, through 'micromutations' and shifts in gene frequency at polygenic loci. In the present section I will briefly examine a mechanism for the establishment of a macromutation, in which the niche concept is central. I present it here simply as something that is possible and improbable—that is, something which would be expected to happen, but only very infrequently. Discussion of the significance of occasional instances of this mechanism, which may be quite profound despite its rarity, can be found in Arthur (1984).

It is perhaps best to start by stressing why many evolutionary biologists have felt the need to invoke at least occasional macromutations, and why a special mechanism other than standard Darwinian selection is 'required' for their establishment at the population level.

The appeal of macromutations lies in their ability to achieve, in a single step, a major phenotypic change that it would be either impossible, or at least slow and difficult, to produce through a gradual accumulation of smaller changes. An example of an evolutionary step which could not under any circumstances be produced gradually is the shift from dextrality to sinistrality that occurred (among other occasions) at the origin of the gastropod family Clausiliidae. (See Arthur, 1984, for a fuller account; other recent important works on the phenomenon of chirality in gastropods are those by Freeman and Lundelius, 1982, and Gould *et al*., 1985.) In many other cases, such as increases or decreases in the number of segments in an insect, there is scope for saltational change—though it could be argued by a neo-Darwinian purist that, for example, a segment could be gradually 'worn away' in an evolutionary sense, until it eventually disappeared.

The difficulty in proposing an evolutionary role for macromutations lies not at the level of the individual organism but at the level of the population. The occurrence of mutations of very large and diverse phenotypic effects is well known, particularly in *Drosophila*, and many of these are 'spontaneous' and so will occur in natural populations and not just in the laboratory. The problem that we encounter lies in what happens after a macromutation occurs.

The problem is essentially that the phenotypically deviant organisms produced by macromutations are usually if not always very unfit compared to the 'wild type' as we can label the typical individual in the population in which our macromutation occurs. This is known both from the theoretical argument of Fisher (1930) which connects increasing magnitude of phenotypic effect of a mutation with increasing probability of being at a selective disadvantage; and from population studies of various major mutations in *Drosophila*, in which the wild-type allele rapidly reaches fixation regardless of starting frequency or ecological conditions such as temperature, humidity, food-source, and so on.

Suppose that macromutants, however the category is delimited (and it has no clear boundaries), are always unfit relative to the wild-type. Is there then any means by which they might spread and take part in a major evolutionary change, rather than being removed from the population, shortly after their occurrence, by selection? The conventional answer, of course, is no; but my purpose here is to suggest otherwise. And, strange as it must seem, I turn not to a non-selective force that can alter gene frequencies, such as genetic drift, but rather to the very mechanism that gives rise to our problem—selection.

At this point I turn to the deceptively simple question: what is selection? Since the phrase 'survival of the fittest' is synonymous to natural selection, our question can equally well be put in the form: what is fitness? The answer that we get to this question depends entirely on what type of evolutionary biologist we approach. Population geneticists normally define the fitness of a genetic variant relative to another, using the cross-product-ratio w described in Chapter 6 or some related measure. I will call this kind of fitness *relative* fitness and the kind of selection based on it Darwinian or w-selection. Evolutionary biologists outside the confines of population genetics, on the other hand, sometimes reject this measure of fitness, though usually without providing us with any clearly-defined substitute. For example, Stearns (1984) states that 'within the broader community of evolutionary biologists, population geneticists are distinguished by their unique conviction that they know what fitness is'.

While I must agree with population geneticists that the Darwinian concept of relative fitness seems appropriate in the vast majority of evolutionary scenarios, I also think that it is profitable to enquire whether there are some conditions under which it is not appropriate. Now it is commonly supposed that relative fitness is the appropriate measure when considering the fate of a genetic variant in relation to another variant in the same population, i.e. one with which it interbreeds. This, if not a statement of the obvious, is certainly a statement of the 'taken-for-granted'. After all, the literature on population genetics is full of studies on the relative fitness of conspecific genetic variants (see Figure 3.2 for example) and notably lacks studies in which a genetic variant of the species has its fitness assessed against an entirely different species. Nevertheless, the idea that interbreeding is the 'necessary and sufficient' condition for relative measurement of fitness is erroneous, and two clues to the correct condition can be found in earlier chapters.

The first clue, in Chapter 2, is that competitive exclusion is analogous to directional selection within a species. The fate of a species in such a context can also be assessed using the cross-product-ratio w, which is then a measure of competitive ability (see Arthur, 1980a). Thus relative measurement of 'fitness' is sometimes appropriate outside a single-species context. The other clue is in Chapter 6, where we saw that the concept of fitness became muddied, in a polymorphic situation, when the different variants had different 'niches' in the sense of different limiting factors. Here, the relative fitness of one variant to the other under any one set of conditions was insufficient to predict what would happen in the two-niche situation.

What these two observations tell us is that given competition but no interbreeding, it is relative fitness that is important; and that given interbreeding but, in a sense, no competition, straightforward relative fitnesses are inadequate. That is, the necessary and sufficient condition for relative measurement of fitness to be appropriate is competition, not interbreeding. Also, I should stress that 'competition' here is used in the broad sense of sharing a limiting factor, not the narrower exploitative sense of sharing a limiting resource.

This conclusion sheds a different light on the fate of a newly-arising macromutation. If it competes with its own wild-type (or, presumably, the wild type of another species), then it will be eliminated in a standard process of w-selection. If, on the other hand, it does not so compete, it may, after all, survive and establish a population. For example, many macromutants in *D. melanogaster* are eliminated in competition with wild-type *D. melanogaster* or with wild-type *D. simulans*, but readily establish a stable, dense population in monoculture.

In such laboratory situations, lack of competition arises through lack of sympatry. In nature, macromutants will always arise where their own wild-type is present—and other competing species will often be present also. Lack of competition can then only arise by isolation in niche terms rather than spatial ones. Macromutants whose niches are altered in such a way that they do not compete with other forms do not undergo w-selection. They still do undergo a form of selection, though, in the sense that some such mutants will establish a population, while others will not. The criterion for success now, however, is not $w_{\text{mutant}} > w_{\text{wild-type}}$, but simply, for the mutant subpopulation itself, the net reproductive rate, $R_o > 1$.

The biggest difficulty for the operation of this process of n-selection—as I have called it (Arthur, 1984) on the basis of the *net* reproductive rate being crucial—lies in the realm of interbreeding. If the macromutant cannot interbreed with the wild-type, its establishment is impeded by the fact that mutations occur singly, or at best a few at a time. If it does interbreed with the wild-type, then it can not become the origin of a new evolutionary line until that reproductive link is somehow broken. There are some ways around these problems, as I have discussed elsewhere (Arthur, 1984). Also, Erwin and Valentine (1984) have made the novel suggestion that 'horizontal' transmission of transposable elements causing macromutations might be achieved by RNA-based viruses, thus removing the 'single mutation' obstacle. Such suggestions are speculative, of course, but they do raise interesting possibilities.

Since the weaker screen of n-selection (compared to that of w-selection) will only operate when new mutants do not compete with any pre-existing wild types, this process would be expected to become rarer as evolution proceeds and fewer areas of unoccupied niche space are available. Because of

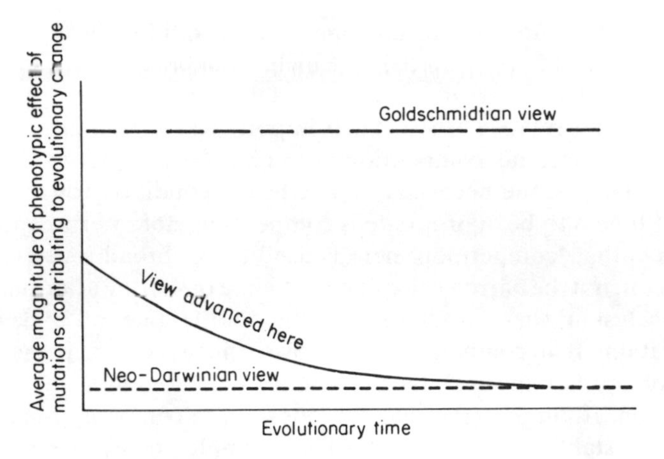

Figure 9.3 Comparison of three views of long-term evolution. (Note that short-term fluctuation around each of the three idealized lines is not shown here)

this effect, an evolutionary process in which occasional radically-altered mutant types are established through n-selection would be expected to differ both from a neo-Darwinian evolutionary process, and from a predominantly saltational one, such as proposed by Goldschmidt (1940), as shown in Figure 9.3. Indeed, there may be 'internal' as well as ecological reasons to expect the curvilinear pattern shown in the figure. In particular, the nature of embryological development may have altered over the course of evolution so that establishment of macromutations by conventional w-selection has also become increasingly improbable over evolutionary time (Erwin and Valentine, 1984). For either or both of these reasons, I suspect that the curve in Figure 9.3 is nearer to the truth than the straight lines of the conventionalists or the early heretics.

Finally, I should stress that the concept of n-selection is yet another version of the main message of this chapter—that major evolutionary shifts are associated, one way or another, with a lack of competition. This is not to say that competition does not sometimes cause evolutionary changes, but rather that such changes tend to be of a more subtle and minor kind than those which are induced by empty niche space.

Chapter 10

Conclusions and Final Remarks

10.1 INTRODUCTION

A particular syndrome, of the following form, can often be found in scientific papers. In the 'results' section, a particular interpretation is made of a particular set of data, though it is made clear that there are plenty of 'ifs and buts' and that alternative explanations are equally viable. In the 'discussion' the alternatives are downgraded from a variety of viewpoints, none of them based on hard data, and made to seem much less likely than the chosen interpretation. Finally, that interpretation appears in the 'abstract' (or summary) as a bald statement. The scientific literature being extensive and scientists' reading time limited, abstracts get read more than main texts, and their 'conclusions' become accepted and gradually embedded in the subsequent literature until, perhaps, the bubble bursts as a result of some new set of data.

Bald 'conclusions', which are not really conclusions, can also creep insidiously into the concluding sections of review articles and books. For example, Abrams (1983) in an otherwise good review, concludes at the end that 'character displacement . . . generally does not result in species being separated by their limiting similarity'. There are simply no grounds for such a sweeping generalization. Abrams may think that character displacement has this (lack of) effect, and some theoretical models may suggest this, but it is certainly not something that can be said to be generally true. On the present evidence from natural communities, we cannot even be sure whether character displacement actually occurs, let alone exactly what amount of divergence of RUFs it will produce; and indeed it can be argued, as noted in Chapter 7, that persistent selection resulting from interspecific competition will only be found where species coexist and so may already be separated by more than their limiting similarity.

In order to avoid perpetuating the syndrome discussed above, I have structured this concluding chapter so as to include separate sections on genuine conclusions (Section 10.2) and what I have called 'concluding speculations' (Section 10.3). The latter are those generalizations which are suggested by at least some models, experiments or field surveys, and which seem to me to be potentially valuable, but which may yet turn out to be false. Unfortunately, but inevitably at this embryonic stage in our understanding of nature, the conclusions apply mostly to simple laboratory systems, the speculations to natural communities.

The remaining two sections of this chapter deal with (a) one of the most central issues in population biology—the concept of limiting factors (Section 10.4)—and (b) the question of how best to employ the term 'niche' in future work (Section 10.5). Although 'merely' a terminological issue, this last one is of considerable importance. The use of 'niche' in previous studies to mean anything from a precise description of the feeding behaviour of a population to some general measure of 'its ecology' has been seriously detrimental to population biology in general, and to competition theory in particular.

10.2 CONCLUSIONS

Simple competitive systems involving organisms such as *Paramecium* and *Drosophila* were discussed in Chapters 2 and 5. What can we conclude about the behaviour of such systems? One obvious fact, perhaps disguised in earlier chapters because of my concentration on stability, is that the vast majority of these simple competitive systems result in the exclusion of one of the competing species—or at least exhibit a clear trend in this direction. This tells us two things. First, most of the simple systems concerned lack a competitive stabilizing mechanism. Second, interspecific differences are usually so great, even between close congeners, that the fate of a two-species (or multi-species) system is never left entirely in the hands of stochastic forces, as may sometimes happen in the analogous situation of competition between genotypes within a species.

It must be stressed that the prevalent tendency towards competitive exclusion in simple laboratory systems does not tell us what the stabilizing mechanism is in more complex, natural systems where a greater proportion of 'competition events' are thought to result in coexistence. Of course, the simple laboratory environments used often drastically curtail the potential for resource partitioning. However, most experiments used only two species, thereby eliminating the possible stabilizing effect of non-transitive competitive abilities. Also, many experiments (such as those of Park (1948, 1954) used very small numbers of founders and thus greatly diminished the prospect of any stabilizing mechanism based on genetic variability (e.g. 'genetic feedback') operating successfully. In general, laboratory studies of competitive exclusion have been based on systems which lack *all* potential stabilizing mechanisms. I stress this point because some authors have concluded otherwise and thought that the promotion of competitive exclusion in systems with a 'homogeneous resource' implies that resource heterogeneity, and associated niche differentiation, is the stabilizing mechanism in nature. For example, MacArthur (1972) states that 'the early bottle experiments on competition . . . proved that environmental heterogeneity is essential for coexistence by showing that in bottles where heterogeneity was absent, coexistence was rare if not impossible'. I happen to think, as MacArthur apparently did, that resource partitioning *is* the predominant cause of natural competitive coexistence; but I do not think that we can conclude this from the prevalence of competitive exclusion in the laboratory.

What is much more informative is the experimental analysis of patterns of resource use, and other ecological parameters, in those few simple laboratory systems which did result in stable coexistence. As we have seen, such analyses are often lacking, such as in the case of Gause's *Paramecium*. However, in other cases, such as the *Drosophila* system described in Chapter 5, analysis of resource use revealed clear niche differentiation. We can conclude from these experiments that resource partitioning can cause stable coexistence in practice. While this result is hardly surprising, it is noteworthy that no experiments have been conducted that clearly demonstrate the efficacy of any of the other stabilizing mechanisms that have been proposed (see Chapter 4) in even a single system. Thus, while the weak version of the mechanistic CEP, which claims that resource partitioning is the *prevalent* cause of competitive coexistence in the real world, seems more likely to be true that the strong version, which claims that it is the *only* such cause, the strong version cannot, at the moment, be rejected.

Let us now turn briefly to the process of intergenotypic, intraspecific competition. Since I have given a treatment of this which is restricted to emphasizing its parallels with competition between species, this is not an appropriate place to attempt any ambitious conclusions. However, certain basic messages are apparent, even from the limited discussions in Chapters 3 and 6. First, rapid competitive elimination of one allele by another is frequently seen in the laboratory, and seen, though much less frequently, in the field. This process parallels competitive exclusion. Second, most enzyme polymorphisms do not show this kind of unstable behaviour and therefore either the different genotypes are so nearly equal in competitive ability that the system is neutrally stable (Kimura, 1983) or they are held in a state of balance by some form of balancing selection. Niche differentiation among genotypes is a potentially widespread form of such balancing selection, acting, as it does, in a frequency-dependent manner. However, there is insufficient evidence to determine the commonness of this kind of balancing selection; and indeed it could still be argued that we do not have any really clear evidence of its operation in any particular system.

It will be apparent that the main contribution of experimental studies in the laboratory concerns *mechanisms*, and many further such studies are needed, despite their current unpopularity. Of course, field studies are needed to tell us what happens in nature, and the increasing use of field experiments in particular is to be commended. However, while a field experiment may realistically be expected to tell us whether two species are in competition in a certain sympatric locality, they usually cannot tell us if there is a stable equilibrium, and if so what is its cause. There is thus a continuing need for detailed laboratory analysis of competitive stabilizing mechanisms to be conducted alongside the field worker's quest for information on the commonness of competition in nature.

10.3 CONCLUDING SPECULATIONS

Are the conclusions described in the previous section applicable to nature? If so, how may they be applied to the more complex scenario of a natural environment? If not, why not? Because exploitative competition is rarely found in nature or because when it is found its dynamics are in some way fundamentally different to those of simple systems? In what way do genetic changes in competing populations modify the competitive process itself? How does competition, or the lack of it, contribute to long-term evolution?

These are just some of the many important questions that niche theorists and other population biologists would like to see answered but which, as yet, defy any clear conclusions. In the absence of such conclusions I will briefly outline some possible generalizations which should serve as a basis for discussion. I will concentrate on the three areas represented in the last three chapters—coevolution, community structure and long-term evolution, in that order. What follows is, essentially, an expansion of the three 'messages' that were identified in the preamble to Part III.

10.3.1 Coevolution

It is impossible to believe that competitively-induced coevolution does not take place in nature. Field experiments, supported by other, less conclusive, data, point to the occurrence of interspecific competition in nature (albeit to an as yet unspecified extent) and to the occurrence of sympatric areas where competing species coexist. All outbreeding organisms exhibit considerable genetic variability in their populations, much of which affects fitness, and some of which affects the 'interspecific competitive ability' component of fitness. Given these conditions, some form of coevolution must occur.

This rather general, but positive view might at first seem to be incompatible with the view—expressed in Chapter 7—that there is little if any solid evidence for the most widely-discussed form of competitive coevolution, namely character displacement. However, this incompatibility is only apparent, and there are at least two reasons why the two views are not contradictory. One possibility, of course, is simply that it is the evidence, and not character displacement itself, that is thin on the ground. I think this is unlikely, but it cannot be ruled out. A more likely possibility is that coevolution often takes forms other than displacement, and indeed that it often has a relatively 'cryptic' effect and does not involve morphological change. Coevolution may take different forms in different cases, especially given a different starting array of genotypes, and may often be de-stabilizing in its effect on coexistence rather than stabilizing (see Arthur and Middlecote, 1984b for an example). Indeed, since coevolution is only likely to occur when there is already coexistence there is, in a sense, more scope for a destabilizing effect. Also, in some cases, only one competing species may have time to evolve in competitive ability before the coexistence collapses. (Whether or not this represents *co*evolution depends on how narrowly the term is defined.) Admittedly, a heterogeneous collection of coevolutionary changes does not connect with niche theory as clearly as the single concept of character displacement does. This may be partly why the idea of displacement has persisted, without much clear evidence for it—but it is not a 'good' reason, and if the choice is between a messy but accurate picture and a clean but false one, we should not hesitate to choose the former.

10.3.2 Community structure

To look for the most direct effects of interspecific competition on community structure, it seems sensible, as noted in Chapter 8, to focus on guilds—and many studies have done precisely this. It seems increasingly clear that many guilds in natural communities are food-limited and 'competitive', and also that many are limited by predators or other factors and do not exhibit exploitative competition at all. It is less clear what the relative abundances of these two types of guild are, or whether there is any pattern in their occurrence—though the suggestion of Hairston *et al*. (1960) that herbivores are usually not food-limited while other trophic types often are limited in this way still deserves serious consideration a quarter century on.

Although natural competitive guilds no doubt exist, the way they are structured by competition has received relatively little intensive study. Too many workers have inferred patterns of resource utilization indirectly from morphological character distributions. The idea of a precisely regular spacing of niches along a resource dimension seems overly simplistic. However, if the mechanistic CEP is generally true, together with some limiting similarity criterion, then some degree of regularity should be detectable. In cases where it is found, I would hazard a guess that it is largely determined by competitive exclusion rather than by character displacement; and moreover that the RUFs of the incoming species have been moulded by all sorts of past evolutionary influences, not just previous competitive ones.

One potential source of competitive structuring of guilds that is not adequately dealt with in niche theory is ranking of competitive abilities. There has been increasing emphasis in the recent literature on asymmetrical competition, where one species is clearly superior to the other (e.g. Wilson, 1975; Lawton and Hassell, 1981, 1984, Persson, 1985). In a sense there are two dimensions to the placing of species in competitive terms: a horizontal axis depicting 'type' and a vertical one depicting 'ability'. Conventional niche theory is obsessed with the former; a more balanced theory of competition would somehow incorporate both—but it is not yet clear how this should be accomplished.

10.3.3 Long-term evolution

In contrast to evolutionary changes caused by interspecific competition, which are relatively minor and have been difficult to document, evolutionary proliferation of species into new, previously vacant niche space is a much more dramatic, and better-evidenced process. Indeed, all 'adaptive radiations', from the very major (e.g. the mammals) to the relatively minor (like Darwin's finches on the Galapagos) can be thought of as examples of precisely this process.

Of course, the adaptive radiation of a large group can no more be completely explained in terms of directional selection at individual loci (as described in Chapter 3) than community structure can be completely described in terms of competitive exclusion. In particular, the nature and frequency of mutations at any one locus is important, as is the poorly understood but crucial network of interactions between different loci and their products that bridges the gap between gene and organism. But these are 'internal' considerations and not appropriately discussed herein.

On the external, ecological side, I reiterate my belief that in the very long term we see, in any major taxon, a shift from major modification to minor modification to stabilization as the group concerned proliferates to fill an originally empty area of ecospace. Whether this view is correct depends in large measure on the time-lag between evolution of a new array of resource organisms and evolution of a new set of consumers. The longer the time-lag the better for the view outlined above. If, on the other hand, consumer evolution could track resource evolution very rapidly, then there would never be any large, vacant extent of niche-space available. The fact that this question can be posed at all, without receiving a clear answer, emphasizes the need for a more fully developed theory of 'community evolution'—a phrase that is intended not to imply the existence of some level of selection above even the species selection of Stanley (1975, 1979), but simply that evolutionary processes, however driven, should be considered in a community context.

10.4 ON LIMITING FACTORS

The concept of a limiting factor is central to the whole sphere of population-biological endeavour, bearing on topics ranging from polymorphism to community structure. It is thus essential that we have some clear idea of what is meant by the term. Unfortunately, like 'niche', 'limiting factor' is used in a variety of ways; and the matter is further complicated by the existence of the related terms 'controlling factor', 'regulating factor' and 'density-dependent factor', with varying degrees of implied synonymy among them.

I do not wish to attempt, here, to propose a 'best usage' of these terms in the (unrealistic) hope that it would become general. But it is necessary to state the actual usage that I will adopt if the following discussion is to have any meaning.

I use 'density-dependent factor' to mean any factor (e.g. predation, competition) that causes natality and/or mortality rates to behave in a density-dependent way. Whether a particular density-dependent factor (e.g. competition for food) actually is effective in producing an equilibrium population size in any particular case is left undefined in the above. I use 'limiting factor' to mean any density-dependent factor that is effective in this sense—i.e. one which causes a population to keep returning towards an equilibrium value after perturbations away from it. Thus the limiting factor category is a subset of the density-dependent one. I will consider the terms 'regulating' and 'controlling factor' to be synonymous with limiting factor, and will not use them.

It should be noted that density-dependent factors can, in a sense, be either too strong or too weak to produce an equilibrium population size. For example, a weak density-dependent effect of competition for food, which could result in a high equilibrium population size, might not do so because of the 'stronger' density-dependent effect of a predator which produces a lower equilibrium. At the other extreme, a very strongly density-dependent effect of competition for food can, in a discrete-growth system (or a continuous one with time-lags) result in 'overcompensation' and, in extreme cases such as some insect populations in the laboratory, cause extinction. Note that 'strength' here refers to the steepness of the slope of per cent mortality with density (or the steepness of the downward slope of per capita natality with density). In using the term 'limiting factor', I am excluding factors which cause these non-equilibrium-producing forms of density dependence.

One obvious implication of the notion of a limiting factor is that variation in the level of that factor—be it food, predation, or whatever—will be followed by variation in equilibrium population size, whereas variation in the level of other factors will not, except where such variation causes a *change* of limiting factor.

Accepting this general picture for the moment, its relation to competition and the niche is clear. Competition and sharing a limiting factor become one and the same thing, as pointed out a long time ago by Williamson (1957). Where the limiting factor is a resource, competition is exploitative; where it is predation, the competition is 'apparent' (Holt, 1977). If species do not share a limiting factor, the reason for their coexistence or lack of it cannot be sought in the realm of competitive stabilizing mechanisms. These apply when such a factor *is* shared.

It is worth briefly examining how the main niche-based stabilizing mechanism—resource partitioning—relates to the limiting factor concept. Taking again the well-worn example of seed size variation and granivorous birds, we can ask the question 'what constitutes the limiting factor for each species?' in a situation where two species have partially overlapping RUFs on the seed size axis. Assuming that it is food (in the form of seeds) that is *generally* limiting, then what limits each species is the supply of seeds in the range that it utilizes. That is, the limiting factor corresponds with the base of the RUF for each species. Competition still equates with 'sharing a limiting factor', provided that sharing is interpreted as implying any (unspecified) degree of overlap. Stable coexistence equates with overlap of limiting factors being incomplete, if we accept the idea of limiting similarity.

What all this says is that there is no conflict between the niche-based theory of exploitative competition and the limiting factor concept provided (a) that this concept is realistic and (b) that in at least some systems populations are limited from 'below' by food (or space) rather than from 'above' by predators. We have examined the evidence for (b) in Chapter 8. However, that discussion, like those of other authors debating the 'upward versus downward limitation' issue (e.g. Pimm, 1980, 1984), did not address (a) and indeed assumed that the limiting factor concept was acceptable. It is this assumption that we now need to query.

In fact, it is quite possible that a population has no single limiting factor. This has been stressed by, among others, Taylor (1984), who illustrated his point with the isocline diagrams reproduced (with some modification) here as Figure 10.1. In Figure 10.1(a) we see a joint stable equilibrium for prey and predator populations (arrowed where the isoclines intersect). It should be noted that the shape of the prey isocline incorporates (by falling off at high prey density) an intraspecific density-dependent effect such as would result from competition for food. In Figure 10.1(b), we see that either supplementing the food supply or removing the predator will result in a shift in the equilibrium prey population size from H_1 to H_2. Thus in this situation there is no single limiting factor despite the fact that the population tends towards a stable equilibrium value.

The same point can be made in a different way. Suppose that we have a population of organisms that can perceive the risk associated with being active while predators are present, and are consequently less often active then than when predators are absent. The equilibrium population size might depend on the amount of resource consumed per individual, which in turn might depend on TR, where T is the activity period and R the resource density. Increasing the resource density or removing the predator in a manipulative field experiment would result in an increase in N. Again, the idea of a single limiting factor simply does not apply.

The main message here seems to be this: as well as assessing how often food, rather than (say) predation, is limiting, we also need to assess the frequency of situations in which the single limiting factor concept is applicable. Indeed, the latter is logically prior to the former. Manipulative experi-

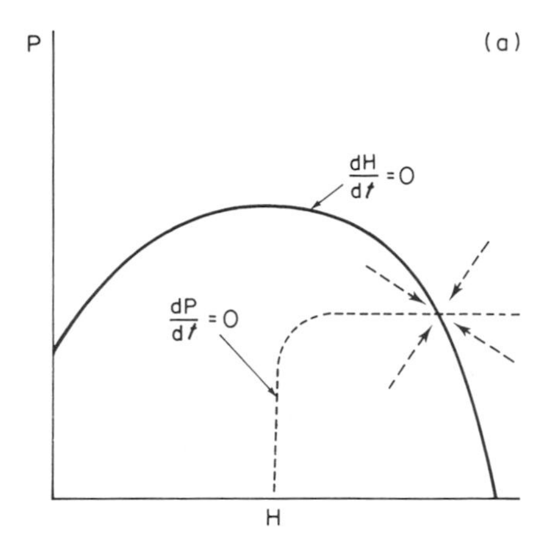

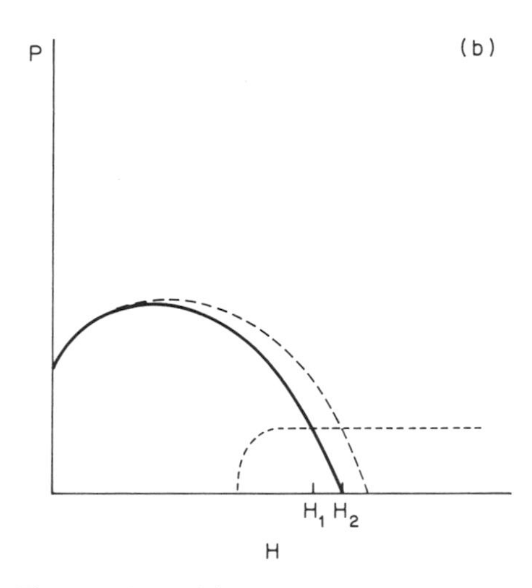

Figure 10.1 (a) Stable equilibrium of consumer (P) and resource (H) populations. (b) Demonstration that either removal of consumer species or supplementing of H's food supply can shift the resource population's equilibrium from H_1 to H_2

ments (in field and laboratory) can be used to address the broader issue identified here—though their design will necessarily be complex and will incorporate modification of at least two factors rather than just one.

10.5 WHAT NICHE FOR THE FUTURE?

At the beginning of this book I gave a selection of quotations designed to reflect the spectrum of emotions—from enthusiasm to exasperation—which the niche concept, in its various forms, has engendered in ecologists. In Chapter 1, I discussed the main versions of the niche conept—those of Elton, Hutchinson and MacArthur. Having finally got to the end of the book, I find myself identifying most closely with the mixed feelings of Root (1967). That is, I do think that the niche is potentially an important concept but also that it is a difficult one and, further, that many of its difficulties arise from its multiplicity—the fact that it is really not one concept, but several.

Faced with a variety of usages, I chose, at the outset, to equate 'niche' with MacArthur's RUF, so that a niche became, as noted in Chapter 1, a property of a particular population–environment combination. In this final section I wish to re-emphasize the main benefit of this choice, and to suggest that future workers investigating the effects of exploitative competition on community structure go along with it.

The benefit is simply this: RUFs are measurable, and ideas like limiting similarity consequently testable. Of course, by narrowing down our concept of the niche and making it testable, we risk niche-based explanations being tested and rejected. So, from another viewpoint, the adoption of the RUF as the niche is disadvantageous because it increases the chances that niche theory will eventually be abandoned as a valid explanation of natural phenomena. However, while my present view is that a refined version of current niche theory will be found to apply to some guilds (but not others) and that it will not be entirely dismissed, I would view its dismissal (on valid experimental grounds) as a greater advance for ecology than the permanent existence of a broader, vaguer and ultimately untestable 'niche theory' based on a version of the niche that is sufficiently fluid to side-step experimental attack. In other words, we are better off having a niche theory that constitutes one of a series of alternative and falsifiable views of the ecological world than having 'niche theory' as a label for a large and heterogeneous collection of views. Having made that choice, we can stop philosophizing and get on with testing the theory.

References

Abrams, P. (1983) The theory of limiting similarity. *Ann. Rev. Ecol. Syst.*, **14**, 359–376.

Allison, A. C. (1955) Aspects of polymorphism in man. *Cold Spring Harbor Symp. Quant. Biol.*, **20**, 239–255.

Andrewartha, H. G., and Birch, L. C. (1954) *The Distribution and Abundance of Animals*. University of Chicago Press, Chicago.

Andrewartha, H. G., and Birch, L. C. (1984) *The Ecological Web: More on the Distribution and Abundance of Animals*. University of Chicago Press, Chicago.

Antonovics, J., Bradshaw, A. D., and Turner, R. G. (1971) Heavy metal tolerance in plants. *Adv. Ecol. Res.*, **7**, 1–85.

Armstrong, R. A., and McGehee, R. (1976a) Coexistence of two competitors on one resource. *J. Theor. Biol.*, **56**, 499–502.

Armstrong, R. A., and McGehee, R. (1976b) Coexistence of species competing for shared resources. *Theor. Pop. Biol.*, **9**, 317–328.

Armstrong, R. A., and McGehee, R. (1980) Competitive exclusion. *Am. Nat.*, **115**, 151–170.

Arthur, W. (1977) *Studies on the relationship between environmental heterogeneity, natural selection and interspecific competition*. Ph.D. Thesis, University of Nottingham.

Arthur, W. (1978) Morph-frequency and coexistence in *Cepaea*. *Heredity*, **41**, 335–346.

Arthur, W. (1980a) Interspecific competition in *Drosophila*. I. Reversal of competitive superiority due to varying concentration of ethanol. *Biol. J. Linn. Soc.*, **13**, 109–118.

Arthur, W. (1980b) Interspecific competition in *Drosophila*. II. Competitive outcome in some 2-resource environments. *Biol. J. Linn. Soc.*, **13**, 119–128.

Arthur, W. (1980c) Further associations between morph-frequency and coexistence in *Cepaea*. *Heredity*, **44**, 417–421.

Arthur, W. (1982a) The evolutionary consequences of interspecific competition. *Adv. Ecol. Res.*, **12**, 127–187.

Arthur, W. (1982b) A critical evaluation of the case for competitive selection in *Cepaea*. *Heredity*, **48**, 407–419.

Arthur, W. (1984) *Mechanisms of Morphological Evolution: A Combined Genetic, Developmental and Ecological Approach*. John Wiley & Sons, Chichester.

Arthur, W. (1986) On the complexity of a simple environment: competition, resource partitioning and facilitation in a 2-species *Drosophila* system. *Phil. Trans. Roy. Soc. B*, **313**, 471–508.

Arthur, W., and Middlecote, J. (1984a) Frequency dependent competitive abilities and differential resource use in competition between *Drosophila hydei* and *D. melanogaster*. *Biol. J. Linn. Soc.*, **23**, 167–176.

Arthur, W., and Middlecote, J. (1984b) Evolution of pupation site and interspecific competitive ability in *Drosophila hydei*. *Biol. J. Linn. Soc.*, **23**, 343–352.

Atkinson, W. D., and Shorrocks, B. (1977) Breeding site specificity in the domestic species of *Drosophila*. *Oecologia*, **29**, 223–232.

Atkinson, W. D., and Shorrocks, B. (1981) Competition on a divided and ephemeral resource: a simulation model. *J. Anim. Ecol.*, **50**, 461–471.

Atkinson, W. D., and Shorrocks, B. (1984) Aggregation of larval Diptera over discrete and ephemeral resources: the implications for coexistence. *Am. Nat.*, **124**, 336–351.

Ayala, F. J. (1966) Reversal of dominance in competing species of *Drosophila*. *Am. Nat.*, **100**, 81–83.

Ayala, F. J. (1969) Experimental invalidation of the principle of competitive exclusion. *Nature*, **224**, 1076–1079.

Ayala, F. J. (1970) Competition, coexistence, and evolution. In: *Essays in Evolution and Genetics in Honor of Theodosius Dobzhansky*, Hecht, M. K., and Steere, W. C. (eds.) North-Holland, Amsterdam.

Ayala, F. J. (1971) Competition between species: frequency dependence. *Science*, **171**, 820–824.

Ayala, F. J., and Campbell, C. A. (1974) Frequency-dependent selection. *Ann. Rev. Ecol. Syst.*, **5**, 115–138.

Bakker, K. (1961) An analysis of factors which determine success in competition for food among larvae of *Drosophila melanogaster*. *Arch. Neerl. Zool.*, **14**, 200–281.

Barker, J. S. F. (1983) Interspecific competition. In: *The Genetics and Biology of* Drosophila, Vol. 3c, Ashburner, M., Carson, H. L., and Thompson, J. N. (eds.) Academic Press, London.

Bellows, T. S., and Hassell, M. P. (1984) Models for interspecific competition in laboratory populations of *Callosobruchus* spp. *J. Anim. Ecol.*, **53**, 831–848.

Birch, L. C. (1957) The meanings of competition. *Am. Nat.*, **91**, 5–18.

Bishop, J. A., and Cook, L. M. (1980) Industrial melanism and the urban environment. *Adv. Ecol. Res.*, **11**, 373–404.

Bishop, J. A., Cook, L. M., and Muggleton, J. (1978) The response of two species of moths to industrialization in Northwest England. *Phil. Trans. Roy. Soc. Lond. B*, **281**, 517–540.

Boag, P. T., and Grant, P. R. (1978) Heritability of external morphology in Darwin's finches. *Nature*, **274**, 793–794.

Bobek, B. (1971) Influence of population density upon rodent production in a deciduous forest. *Ann. Zool. Fennici*, **8**, 137–144.

Bradshaw, A. D. (1984) Adaptation of plants to soils containing toxic metals—a test for conceit. In: *Origins and Development of Adaptation*, Ciba Foundation Symposium 102, Pitman, London.

Brown, J. H., and Davidson, D. W. (1977) Competition between seed-eating rodents and ants in desert ecosystems. *Science*, **196**, 880–882.

Brown, W. L., and Wilson, E. O. (1956) Character displacement. *Syst. Zool.*, **5**, 49–64.

Bulmer, M. G. (1974) Density-dependent selection and character displacement. *Am. Nat.*, **108**, 45–58.

Buss, L. W., and Jackson, J. B. C. (1979) Competitive networks: nontransitive competitive relationships in cryptic coral reef environments. *Am. Nat.*, **113**, 223–234.

Cain, A. J., and Sheppard, P. M. (1957) Some breeding experiments with *Cepaea nemoralis* (L.). *J. Genet.*, **55**, 195–199.

Cameron, R. A. D., and Carter, M. A. (1979) Intra- and inter-specific effects of population density on growth and activity in some helicid land snails (Gastropoda: Pulmonata). *J. Anim. Ecol.*, **48**, 237–246.

Campbell, R. N. (1971) The growth of brown trout *Salmo trutta* L. in northern

Scottish lochs with special reference to the improvement of fisheries. *J. Fish. Biol.*, **3**, 1–28.
Carson, H. L., Hardy, D. E., Spieth, H. T., and Stone, W. S. (1970) The evolutionary biology of the Hawaiian Drosophilidae. In: *Essays in Evolution and Genetics in Honor of Theodosius Dobzhansky*, Hecht, M. K., and Steere, W. C. (eds.) North-Holland, Amsterdam.
Carter, M. A., Jeffrey, R. C. V., and Williamson, P. (1979) Food overlap in coexisting populations of the land snails *Cepaea nemoralis* (L.) and *Cepaea hortensis* (Mull.). *Biol. J. Linn. Soc.*, **11**, 169–176.
Case, T. J., and Gilpin, M. E. (1974) Interference competition and niche theory. *Proc. Natl. Acad. Sci. USA*, **71**, 3073–3077.
Clarke, B. (1962) Balanced polymorphism and the diversity of sympatric species. In: *Taxonomy and geography*, Nichols, D. (ed.) Systematics Association, Oxford.
Clarke, B. (1976) The ecological genetics of host-parasite relationships. In: *Genetic Aspects of Host–Parasite Relationships*, Taylor, A. E. R., and Muller, R. (eds.) Symposia of the British Society for Parasitology, Vol. 14. Blackwell, Oxford.
Clarke, B., and Allendorf, F. W. (1979) Frequency-dependent selection due to kinetic differences between allozymes. *Nature*, **279**, 732–734.
Clarke, B., Arthur, W., Horsley, D. T., and Parkin, D. T. (1978) Genetic variation and natural selection in pulmonate molluscs. In: *Pulmonates, Vol. 2A. Systematics, Evolution and Ecology*, Fretter, V., and Peake, J. (eds.) Academic Press, London.
Clarke, C. A., Mani, G. S., and Wynne, G. (1985) Evolution in reverse: clean air and the peppered moth. *Biol. J. Linn. Soc.*, **26**, 189–199.
Cohen, J. E. (1978) *Food Webs and Niche Space*. Princeton University Press, Princeton, New Jersey.
Cole, L. C. (1960) Competitive exclusion. *Science*, **132**, 348–349.
Colwell, R. K., and Winkler, D. W. (1984) A null model for null models in biogeography. In: *Ecological Communities: Conceptual Issues and the Evidence*, Strong, D. R., Simberloff, D., Abele, L. G., and Thistle, A. B. (eds.) Princeton University Press, Princeton, New Jersey.
Connell, J. H. (1980) Diversity and the coevolution of competitors, or the ghost of competition past. *Oikos*, **35**, 131–138.
Connell, J. H. (1983) On the prevalence and relative importance of interspecific competition: evidence from field experiments. *Am. Nat.*, **122**, 661–696.
Connor, E. F., and Simberloff, D. (1978) Species number and compositional similarity of the Galapagos flora and avifauna. *Ecol. Monogr.*, **48**, 219–248.
Connor, E., and Simberloff, D. (1979) The assembly of species communities: chance or competition? *Ecology*, **60**, 1132–1140.
Cook, L. M. (1971) *Coefficients of Natural Selection*. Hutchinson, London.
Cook, L. M. (1983) Detection of frequency-dependent selection in sequences of data. *Heredity*, **50**, 321–322.
Cowie, R. H., and Jones, J. S. (1987) Ecological interactions between the land snails *Cepaea nemoralis* and *Cepaea hortensis*: invasion, competition, but no niche displacement. *Functional Ecol.*, (in press).
Cox, C. B., and Moore, P. D. (1985) *Biogeography: An Ecological and Evolutionary Approach*, (4th edn) Blackwell, Oxford.
Crombie, A. C. (1945) On competition between different species of graminivorous insects. *Proc. R. Soc. B.*, **132**, 362–395.
Crombie, A. C. (1946) Further experiments on insect competition. *Proc. Roy. Soc. Lond. B*, **133**, 76–109.
Crozier, R. H. (1974) Niche shape and genetic aspects of character displacement. *Amer. Zool.*, **14**, 1151–1157.
Darwin, C. (1859) *On the Origin of Species by Means of Natural Selection, or the Preservation of Favoured Races in the Struggle for Life*. John Murray, London.

Day, T. H., Hillier, P. C., and Clarke, B. (1974) Properties of genetically polymorphic isozymes of alcohol dehydrogenase in *Drosophila melanogaster*. *Biochem. Genet.*, **11**, 141–153.

de Jong, G. (1976) A model of competition for food. I. Frequency-dependent viabilities. *Am. Nat.*, **110**, 1013–1027.

de Vries, H. (1905) *Species and Varieties, Their Origin by Mutation*. Open Court Publishing Co., Chicago.

Dempster, J. P. (1982) The ecology of the Cinnabar Moth, *Tyria jacobaeae* L. (Lepidoptera: Arctiidae). *Adv. Ecol. Res.*, **12**, 1–36.

Dobzhansky, T. (1970) *Genetics of the Evolutionary Process*. Columbia University Press, New York.

Dunham, A. E., Smith, G. R., and Taylor, J. N. (1979) Evidence for ecological character displacement in western American catostomid fishes. *Evolution*, **33**, 877–896.

Eldredge, N., and Gould, S. J. (1972) Punctuated equilibria: an alternative to phyletic gradualism. In: *Models in Paleobiology*, Schopf, T. J. M. (ed.) Freeman, San Francisco.

Elton, C. S. (1927) *Animal Ecology*. Sidgwick & Jackson, London.

Elton, C. S. (1958) *The Ecology of Invasions by Animals and Plants*. Methuen, London.

Emlen, J. M. (1973) *Ecology: An Evolutionary Approach*. Addison-Wesley, Reading, Mass.

Ennos, R. A. (1983) Maintenance of genetic variation in plant populations. *Evol. Biol.*, **16**, 129–155.

Erwin, D. H., and Valentine, J. W. (1984) 'Hopeful monsters', transposons, and Metazoan radiation. *Proc. Nat. Acad. Sci. USA*, **81**, 5482–5483.

Felsenstein, J. (1976) The theoretical population genetics of variable selection and migration. *Ann. Rev. Genet.*, **10**, 253–280.

Fenchel, T. (1975) Character displacement and coexistence in mud snails (Hydrobiidae). *Oecologia*, **20**, 19–32.

Fisher, R. A. (1930) *The Genetical Theory of Natural Selection*. Clarendon Press, Oxford.

Ford, E. B. (1971) *Ecological Genetics*, (3rd edn) Chapman & Hall, London.

Freeman, G., and Lundelius, J. W. (1982) The developmental genetics of dextrality and sinistrality in the gastropod *Lymnaea peregra. Wilhelm Roux's Archives*, **191**, 69–83.

Gause, G. F. (1932) Experimental studies on the struggle for existence. I. Mixed population of two species of yeast. *J. Exp. Biol.*, **9**, 389–402.

Gause, G. F. (1934) *The Struggle for Existence*. Williams & Wilkins, Baltimore.

Gause, G. F. (1935) Vérifications expérimentales de la théorie mathématique de la lutte pour la vie. *Act. Sci. et. Ind.* (**277**).

Gause, G. F. (1936) The principles of biocoenology. *Q. Rev. Biol.*, **11**, 320–336.

Gause, G. F. (1937) Experimental populations of microscopic organisms. *Ecology*, **18**, 173–179.

Gause, G. F., and Witt, A. A. (1935) Behaviour of mixed populations and the problem of natural selection. *Am. Nat.*, **69**, 596–609.

Gause, G. F., Natsukova, O. K., and Alpatov, W. W. (1934) The influence of biologically conditioned media on the growth of a mixed population of *Paramecium caudatum* and *P. aurelia*. *J. Anim. Ecol.*, **3**, 222–230.

Gilbert, O., Reynoldson, T. B., and Hobart, J. (1952) Gause's hypothesis: an examination. *J. Anim. Ecol.*, **21**, 310–312.

Gilpin, M. E. (1975) Limit cycles in competition communities. *Am. Nat.*, **109**, 51–60.

Gilpin, M. E., and Diamond, J. M. (1984) Are species co-occurrences on islands non-random, and are null hypotheses useful in community ecology? In: *Ecological Communities: Conceptual Issues and the Evidence*, Strong, D. R., Simberloff, D.,

Abele, L. G., and Thistle, A. B. (eds.) Princeton University Press, Princeton, New Jersey.
Goldschmidt, R. (1940) *The Material Basis of Evolution*. Yale University Press, New Haven, Connecticut.
Gosling, E. (1980) Gene frequency changes and adaptation in marine cockles. *Nature*, **286**, 601–602.
Gould, S. J. (1977) Eternal metaphors of palaeontology. In: *Patterns of Evolution*, Hallam, A. (ed). Elsevier, Amsterdam.
Gould, S. J. (1982) The meaning of punctuated equilibrium and its role in validating a hierarchical approach to macroevolution. In: *Perspectives on Evolution*, Milkman, R. (ed.) Sinauer, Sunderland, Mass.
Gould, S. J., and Eldredge, N. (1977) Punctuated equilibria: the tempo and mode of evolution reconsidered. *Paleobiology*, **3**, 115–151.
Gould, S. J., Young, N. D., and Kasson, B. (1985) The consequences of being different: sinistral coiling in *Cerion*. *Evolution*, **39**, 1364–1379.
Grant, P. R. (1972) Convergent and divergent character displacement. *Biol. J. Linn. Soc.*, **4**, 39–68.
Grant, P. R. (1975) The classical case of character displacement. In: *Evolutionary Biology*, Vol. 8, Dobzhansky, T., Hecht, M. K., and Steere, W. C. (eds.) Plenum Press, New York.
Grant, P. R. (1986) *Ecology and Evolution of Darwin's Finches*. Princeton University Press, Princeton, New Jersey.
Grant, V. (1981) *Plant Speciation*, (2nd edn). Columbia University Press, New York.
Grimaldi, D., and Jaenike, J. (1984) Competition in natural populations of mycophagous *Drosophila*. *Ecology*, **65**, 1113–1120.
Grinnell, J. (1914) An account of the mammals and birds of the Lower Colorado Valley. *Univ. Calif. Publ. Zool.*, **12**, 51–294.
Grinnell, J. (1917) The niche-relationships of the California thrasher. *Auk*, **34**, 427–433.
Grinnell, J. (1924) Geography and evolution. *Ecology*, **5**, 225–229.
Haigh, J., and Maynard Smith, J. (1972) Can there be more predators than prey? *Theor. Pop. Biol.*, **3**, 290–299.
Hairston, N. G. (1985) The interpretation of experiments on interspecific competition. *Am. Nat.*, **125**, 321–325.
Hairston, N. G., Smith, F. E., and Slobodkin, L. B. (1960) Community structure, population control and competition. *Am. Nat.*, **94**, 421–425.
Haldane, J. B. S. and Jayakar, S. D. (1963) Polymorphism due to selection of varying direction. *J. Genet.*, **58**, 237–242.
Hardin, G. (1960) The competitive exclusion principle. *Science*, **131**, 1291–1297.
Harris, H. (1966) Enzyme polymorphisms in man. *Proc. Roy. Soc. B*, **164**, 298–310.
Harris, M. P. (1973a) *Coreba flaveola* and the Geospizinae. *Bull. Br. Orn. Club*, **92**, 164–168.
Harris, M. P. (1973b) The Galapagos avifauna. *Condor*, **75**, 265–278.
Harvey, P. H., Colwell, R. K., Silvertown, J. W., and May, R. M. (1983) Null models in ecology. *Ann. Rev. Ecol. Syst.*, **14**, 189–211.
Haskell, E. F. (1947) A natural classification of societies. *N.Y. Acad. Sci. Trans., Series 2*, **9**, 186–196.
Heath, D. J. (1975) Colour, sunlight and internal temperatures in the land snail *Cepaea nemoralis* (L.). *Oecologia*, **19**, 29–38.
Hedrick, P. W. (1985) *Genetics of Populations*. Jones & Bartlett, Portola Valley, California.
Hedrick, P. W. (1986) Genetic polymorphism in heterogeneous environments: a decade later. *Ann. Rev. Ecol. Syst.*, **17**, 535–566.
Hedrick, P. W., Ginevan, M. E., and Ewing, E. P. (1976) Genetic polymorphism in heterogeneous environments. *Ann. Rev. Ecol. Syst.*, **7**, 1–32.

Holt, R. D. (1977) Predation, apparent competition, and the structure of prey communities. *Theoretical Population Biology*, **12**, 197–229.
Holt, R. D. (1984) Spatial heterogeneity, indirect interactions, and the coexistence of prey species. *Am. Nat.*, **124**, 377–406.
Hubby, J. L., and Lewontin, R. C. (1966) A molecular approach to the study of genic heterozygosity in natural populations. I. The number of alleles at different loci in *Drosophila pseudoobscura*. *Genetics*, **54**, 577–594.
Hutchinson, G. E. (1957) Concluding remarks. *Cold Spring Harb. Symp. Quant. Biol.*, **22**, 415–427.
Hutchinson, G. E. (1959) Homage to Santa Rosalia, or why are there so many kinds of animals? *Am. Nat.*, **93**, 145–159.
Hutchinson, G. E. (1965) *The Ecological Theater and the Evolutionary Play*. Yale University Press, New Haven.
Hutchinson, G. E. (1978) *An Introduction to Population Ecology*. Yale University Press, New Haven.
Jackson, J. B. C. (1979) Overgrowth competition between encrusting cheilostome ectoprocts in a Jamaican cryptic reef environment. *J. Anim. Ecol.*, **48**, 805–823.
Jackson, J. B. C., and Buss, L. W. (1975) Allelopathy and spatial competition among coral reef invertebrates. *Proc. Nat. Acad. Sci. USA*, **72**, 5160–5163.
Jacsic, F. M. (1981) Abuse and misuse of the term 'guild' in ecological studies. *Oikos*, **37**, 397–400.
Jeffries, M. J., and Lawton, J. H. (1984) Enemy free space and the structure of ecological communities. *Biol. J. Linn. Soc.*, **23**, 269–286.
Johnson, R. H. (1910) Determinate evolution in the color-pattern of the lady-beetles. *Carnegie Institution of Washington Publication no. 122*.
Jones, J. S. (1982) Genetic differences in individual behaviour associated with shell polymorphism in the snail *Cepaea nemoralis*. *Nature*, **298**, 749–750.
Jones, J. S., and Probert, R. F. (1980) Habitat selection maintains a deleterious allele in a heterogeneous environment. *Nature*, **287**, 632–633.
Jones, J. S., Leith, B. H., and Rawlings, P. (1977) Polymorphism in *Cepaea*: A problem with too many solutions? *Ann. Rev. Ecol. Syst.*, **8**, 109–143.
Karr, J. R., and James, F. C. (1975) Ecomorphological configurations and convergent evolution. In: *Ecology and Evolution of Communities*, Cody, M. L. and Diamond, J. M. (eds.) Harvard University Press, Cambridge, Mass.
Keller, B. D. (1983) Coexistence of sea urchins in seagrass meadows: an experimental analysis of competition and predation. *Ecology*, **64**, 1581–1598.
Kimura, M. (1968) Genetic variability maintained in a finite population due to production of neutral and nearly neutral isoalleles. *Genet. Res.*, **11**, 247–269.
Kimura, M. (1983) *The Neutral Theory of Molecular Evolution*. Cambridge University Press, Cambridge.
Koch, A. L. (1974a) Coexistence resulting from an alternation of density dependent and density independent growth. *J. Theor. Biol.*, **44**, 373–386.
Koch, A. L. (1974b) Competitive coexistence of two predators utilizing the same prey under constant environmental conditions. *J. Theor. Biol.*, **44**, 387–395.
Kohn, A. J. (1959) The ecology of *Conus* in Hawaii. *Ecol. Monogr.*, **29**, 47–90.
Kojima, K., and Schaffer, H. E. (1967) Survival processes of linked mutant genes. *Evolution*, **21**, 518–531.
Krebs, C. J. (1985) *Ecology: The Experimental Analysis of Distribution and Abundance*, (3rd edn). Harper & Row, New York.
Lack, D. (1947) *Darwin's Finches: An Essay on the General Biological Theory of Evolution*. Cambridge University Press, Cambridge. (Reprinted 1968 by Peter Smith, Gloucester, Massachusetts).
Lack, D. (1971) *Ecological Isolation in Birds*. Blackwell, Oxford.

Lawlor, R., and Maynard Smith, J. (1976) The coevolution and stability of competing species. *Am. Nat.*, **110**, 79–99.

Lawton, J. H. (1976) The structure of the arthropod community of bracken. *Bot. J. Linn. Soc.*, **73**, 187–216.

Lawton, J. H. (1984) Non-competitive populations, non-convergent communities and vacant niches: the herbivores of bracken. In: *Ecological Communities: Conceptual Issues and the Evidence*. Strong, D. R., Simberloff, D., Abele, L. G., and Thistle, A. B. (eds.) Princeton University Press, Princeton, New Jersey.

Lawton, J. H., and Hassell, M. P. (1981) Asymmetrical competition in insects. *Nature*, **289**, 793–795.

Lawton, J. H., and Hassell, M. P. (1984) Interspecific competition in insects. In: *Ecological Entomology*, Huffaker, C. B., and Rabb, R. L. (eds.) John Wiley & Sons, Chichester and New York.

Levene, H. (1953) Genetic equilibrium when more than one ecological niche is available. *Am. Nat.*, **87**, 331–333.

Levin, B. R. (1971) The operation of selection in situations of interspecific competition. *Evolution*, **25**, 249–264.

Levin, B. R. (1972) Coexistence of two asexual strains on a single resource. *Science*, **175**, 1272–1274.

Levins, R., and MacArthur, R. H. (1966) The maintenance of genetic polymorphism in a spatially heterogeneous environment: variations on a theme by Howard Levene. *Am. Nat.*, **100**, 585–589.

Lewontin, R. C. (1974) *The Genetic Basis of Evolutionary Change*. Columbia University Press, New York.

Lewontin, R. C., and Hubby, J. L. (1966) A molecular approach to the study of genic heterozygosity in natural populations. II. Amount of variation and degree of heterozygosity in natural populations of *Drosophila pseudoobscura*. *Genetics*, **54**, 595–609.

Lister, B. C. (1976a) The nature of niche expansion in West Indian *Anolis* lizards. I: Ecological consequences of reduced competition. *Evolution*, **30**, 659–676.

Lister, B. C. (1976b) The nature of niche expansion in West Indian *Anolis* lizards. II: Evolutionary components. *Evolution*, **30**, 677–692.

Lloyd, H. G. (1962) Squirrels in England and Wales 1959. *J. Anim. Ecol.*, **31**, 157–165.

Lloyd, H. G. (1983) Past and present distribution of red and grey squirrels. *Mammal Review*, **13**, 69–80.

Loeschcke, V. (1984) The interplay between genetic composition, species number, and population sizes under exploitative competition. In: *Population Biology and Evolution*, Wöhrmann, K., and Loeschcke, V. (eds.) Springer-Verlag, Berlin.

Lotka, A. J. (1925) *Elements of Physical Biology*. Williams & Wilkins, Baltimore.

MacArthur, R. H. (1958) Population ecology of some warblers of Northeastern coniferous forests. *Ecology*, **39**, 599–619.

MacArthur, R. H. (1968) The theory of the niche. In: *Population Biology and Evolution*, Lewontin, R. C. (ed.) Syracuse University Press, New York.

MacArthur, R. H. (1970) Species packing and competitive equilibrium among many species. *Theor. Pop. Biol.*, **1**, 1–11.

MacArthur, R. H. (1972) *Geographical Ecology: Patterns in the Distribution of Species*. Harper & Row, New York.

MacArthur, R. H., and Levins, R. (1964) Competition, habitat selection, and character displacement in a patchy environment. *Proc. Nat. Acad. Sci. USA*, **51**, 1207–1210.

MacArthur, R. H., and Levins, R. (1967) The limiting similarity, convergence, and divergence of coexisting species. *Am. Nat.*, **101**, 377–385.

MacArthur, R. H., and MacArthur, J. W. (1961) On bird species diversity. *Ecology*, **42**, 594–598.

McMurtie, R. (1976) On the limit to niche overlap for nonuniform niches. *Theor. Pop. Biol.*, **10**, 96–107.

Macfadyen, A. (1963) *Animal Ecology: Aims and Methods*, (2nd edn). Pitman, London.

Malmquist, M. G. (1985) Character displacement and biogeography of the pygmy shrew in northern Europe. *Ecology*, **66**, 372–377.

Margalef, R. (1968) *Perspectives in Ecological Theory*. University of Chicago Press, Chicago.

May, R. M. (1974a) On the theory of niche overlap. *Theor. Pop. Biol.*, **5**, 297–332.

May, R. M. (1974b) *Stability and Complexity in Model Ecosystems*, (2nd edn). Princeton University Press, Princeton, New Jersey.

May, R. M., and MacArthur, R. H. (1972) Niche overlap as a function of environmental variability. *Proc. Nat. Acad. Sci. USA*, **69**, 1109–1113.

Maynard Smith, J. (1966) Sympatric speciation. *Am. Nat.*, **100**, 637–650.

Maynard Smith, J. (1970) Genetic polymorphism in a varied environment. *Am. Nat.*, **104**, 487–490.

Maynard Smith, J. (1972) *On Evolution*. Edinburgh University Press, Edinburgh.

Maynard Smith, J. (1974) *Models in Ecology*. Cambridge University Press, Cambridge.

Maynard Smith, J. (1978) *The Evolution of Sex*. Cambridge University Press, Cambridge.

Mayr, E. (1942) *Systematics and the Origin of Species*. Columbia University Press, New York.

Mayr, E. (1954) Change of genetic environment and evolution. In: *Evolution as a Process*, Huxley, J., Hardy, A. C., and Ford, E. B. (eds.) Allen & Unwin, London.

Mayr, E. (1963) *Animal Species and Evolution*. Harvard University Press, Cambridge, Massachusetts.

Mertz, D. B., and McCauley, D. E. (1982) The domain of laboratory ecology. In: *Conceptual Issues in Ecology*, Saarinen, E. (ed.) Reidel, Dordrecht, Holland.

Middleton, A. D. (1931) *The Grey Squirrel. The Introduction and Spread of the American Grey Squirrel in the British Isles, its Habitat, Food, and Relations with the Native Fauna of the Country*. Sidgwick & Jackson, London.

Middleton, A. D. (1935) The distribution of the grey squirrel (*Sciurus carolinensis*) in Great Britain in 1935. *J. Anim. Ecol.*, **4**, 274–276.

Milligan, B. G. (1985) Evolutionary divergence and character displacement in two phenotypically-variable competing species. *Evolution*, **39**, 1207–1222.

Milne, A. (1961) Definition of competition among animals. *Symp. Soc. Exp. Biol.*, **15**, 40–61.

Moore, J. A. (1952a) Competition between *Drosophila melanogaster* and *Drosophila simulans*. I. Population cage experiments. *Evolution*, **6**, 407–420.

Moore, J. A. (1952b) Competition between *Drosophila melanogaster* and *Drosophila simulans*. II. The improvement of competitive ability through selection. *Proc. Nat. Acad. Sci. USA*, **38**, 813–817.

Murdoch, W. W. (1969) Switching in general predators: experiments on predator specificity and stability of prey populations. *Ecol. Monogr.*, **39**, 335–354.

Murphy, P. G. (1976) Electrophoretic evidence that selection reduces ecological overlap in marine limpets. *Nature*, **261**, 228–230.

Murray, J. (1963) The inheritance of some characters in *Cepaea hortensis* and *Cepaea nemoralis* (Gastropoda). *Genetics*, **48**, 605–615.

Murray, J. (1972) *Genetic Diversity and Natural Selection*. Oliver & Boyd, Edinburgh.

Nicholson, A. J. (1933) The balance of animal populations. *J. Anim. Ecol.*, **2**, 132–178.

Nisbet, R. M., and Gurney, W. S. C. (1982) *Modelling Fluctuating Populations*. John Wiley & Sons, Chichester.
Odum, E. P. (1971) *Fundamentals of Ecology*, (3rd edn). Saunders, Philadelphia.
Paine, R. T. (1966) Food web complexity and species diversity. *Am. Nat.*, **100**, 65–75.
Park, T. (1948) Experimental studies of interspecies competition. I. Competition between populations of the flour beetles, *Tribolium confusum* and *Tribolium castaneum* Herbst. *Ecol. Monogr.*, **18**, 265–307.
Park, T. (1954) Experimental studies of interspecies competition. II. Temperature, humidity, and competition in two species of *Tribolium. Physiol. Zool.*, **27**, 177–238.
Park, T. (1957) Experimental studies of interspecies competition. III. Relation of initial species proportion to competitive outcome in populations of *Tribolium. Physiol. Zool.*, **30**, 22–40.
Park, T. (1962) Beetles, competition, and populations. *Science*, **138**, 1369–1375.
Pask, G. (1961) *An Approach to Cybernetics*. Hutchinson, London.
Pearson, D. L., and Knisley, C. B. (1985) Evidence for food as a limiting resource in the life of tiger beetles (Coleoptera: Cicindelidae). *Oikos*, **45**, 161–168.
Pease, C. M. (1984) On the evolutionary reversal of competitive dominance. *Evolution*, **38**, 1099–1115.
Persson, L. (1985) Asymmetrical competition: are larger animals competitively superior? *Am. Nat.*, **126**, 261–266.
Petit, C. (1968) Le rôle des valeurs sélectives variables dans le maintien du polymorphisme. *Bull. Soc. Zool. Fr.*, **93**, 187–208.
Pianka, E. R. (1966) Latitudinal gradients in species diversity: A review of concepts. *Am. Nat.*, **100**, 33–46.
Pianka, E. R. (1975) Niche relations of desert lizards. In: *Ecology and Evolution of Communities*, Cody, M. L., and Diamond, J. M. (eds.) Harvard University Press, Cambridge, Massachussetts.
Pianka, E. R. (1981) Competition and niche theory. In: *Theoretical Ecology, Principles and Applications*, second edition, May, R. M. (ed.), Blackwell, Oxford.
Pianka, E. R. (1983) *Evolutionary Ecology*, (3rd edn). Harper & Row, New York.
Pimentel, D., Feinberg, E. H., Wood, P. W., and Hayes, J. T. (1965) Selection, spatial distribution and the coexistence of competing fly species. *Am. Nat.*, **99**, 97–109.
Pimm, S. L. (1980) Food web design and the effect of species deletion. *Oikos*, **35**, 139–149.
Pimm, S. L. (1982) *Food Webs*. Chapman & Hall, London.
Pimm, S. L. (1984) The complexity and stability of ecosystems. *Nature*, **307**, 321–326.
Pimm, S. L., and Lawton, J. H. (1977) Number of trophic levels in ecological communities. *Nature*, **268**, 329–331.
Pinsker, W. (1981) MDH-polymorphism in *Drosophila subobscura*: I. Selection and hitch-hiking in laboratory populations. *Theor. Appl. Genet.*, **60**, 107–112.
Pontin, A. J. (1982) *Competition and Coexistence of Species*. Pitman, London.
Prakash, S. (1972) Origin of reproductive isolation in the absence of apparent genic differentiation in a geographical isolate of *Drosophila pseudoobscura*. *Genetics*, **72**, 143–155.
Raff, R. A., and Kaufman, T. C. (1983) *Embryos, Genes, and Evolution: The Developmental-Genetic Basis of Evolutionary Change*. Macmillan, New York.
Reynolds, J. C. (1985) Details of the geographic replacement of the red squirrel (*Sciurus vulgaris*) by the grey squirrel (*Sciurus carolinensis*) in Eastern England. *J. Anim. Ecol.*, **54**, 149–162.
Root, R. B. (1967) The niche exploitation pattern of the blue-gray gnatchatcher. *Ecol. Monogr.*, **37**, 317–350.
Roughgarden, J. (1972) Evolution of niche width. *Am. Nat.*, **106**, 683–718.
Roughgarden, J. (1976) Resource partitioning among competing species—a coevolutionary approach. *Theor. Pop. Biol.*, **9**, 388–424.

Sawicki, R. M., and Denholm, I. (1984) Adaptation of insects to insecticides. In: *Origins and Development of Adaptation*, Ciba Foundation Symposium 102, Pitman, London.
Schluter, D., and Grant, P. R. (1984) Determinants of morphological patterns in communities of Darwin's finches. *Am. Nat.*, **123**, 175–196.
Schluter, D., Price, T. D., and Grant, P. R. (1985) Ecological character displacement in Darwin's finches. *Science*, **227**, 1056–1059.
Schoener, T. W. (1968) The *Anolis* lizards of Bimini: resource partitioning in a complex fauna. *Ecology*, **49**, 704–726.
Schoener, T. W. (1974) Resource partitioning in ecological communities. *Science*, **185**, 27–39.
Schoener, T. W. (1983) Field experiments on interspecific competition. *Am. Nat.*, **122**, 240–285.
Schoener, T. W. (1985) Some comments on Connell's and my reviews of field experiments on interspecific competition. *Am. Nat.*, **125**, 730–740.
Seifert, R. P. (1984) Does competition structure communities? Field studies on neotropical *Heliconia* insect communities. In: *Ecological Communities: Conceptual Issues and the Evidence*, eds. Strong, D. R., Simberloff, D., Abele, L. B., and Thistle, A. B. (eds.) Princeton University Press, Princeton, New Jersey.
Shorrocks, B., Atkinson, W. D., and Charlesworth, P. (1979) Competition on a divided and ephemeral resource. *J. Anim. Ecol.*, **48**, 899–908.
Shorten, M. (1953) Notes on the distribution of the grey squirrel (*Sciurus carolinensis*) and the red squirrel (*Sciurus vulgaris leucourus*). *J. Anim. Ecol.*, **22**, 134–140.
Shorten, M. (1954) *Squirrels*. Collins, London.
Silander, J. A. (1979) Microevolution and clone structure in *Spartina patens*. *Science*, **203**, 658–660.
Silander, J. A., and Antonovics, J. (1979) The genetic basis of the ecological amplitude of *Spartina patens*. I. Morphometric and physiological traits. *Evolution*, **33**, 1114–1127.
Simpson, G. G. (1944) *Tempo and Mode in Evolution*. Columbia University Press, New York.
Sinclair, A. R. E. (1985) Does interspecific competition or predation shape the African ungulate community? *J. Anim. Ecol.*, **54**, 899–918.
Slatkin, M. (1980) Ecological character displacement. *Ecology*, **61**, 163–177.
Slobodkin, L. B. (1961) *Growth and Regulation of Animal Populations*. Holt, Rinehart & Winston, New York.
Slobodkin, L. B., Smith, F. E., and Hairston, N. G. (1967) Regulation in terrestrial ecosystems, and the implied balance of nature. *Am. Nat.*, **101**, 109–124.
Stanley, S. M. (1975) A theory of evolution above the species level. *Proc. Nat. Acad. Sci. USA*, **72**, 646–650.
Stanley, S. M. (1979) *Macroevolution: Pattern and Process*. Freeman, San Francisco.
Stearns, S. C. (1984) Models in evolutionary ecology. In: *Population Biology and Evolution*, eds. Wöhrmann, K. and Loeschcke, V. (eds.) Springer-Verlag, Berlin.
Stewart, F. M., and Levin, B. R. (1973) Partitioning of resources and the outcome of interspecific competition: a model and some general considerations. *Am. Nat.*, **107**, 171–198.
Strobeck, C. (1973) N species competition. *Ecology*, **54**, 650–654.
Strobeck, C. (1974) Sufficient conditions for polymorphism with n niches and m mating groups. *Am. Nat.*, **108**, 152–156.
Strong, D. R. (1984) Exorcising the ghost of competition past: phytophagous insects. In: *Ecological Communities: Conceptual Issues and the Evidence*, Strong, D. R., Simberloff, D., Abele, L. G., and Thistle, A. B. (eds.) Princeton University Press, Princeton, New Jersey.
Strong, D. R., Szyska, L. A., and Simberloff, D. (1979) Tests of community-wide character displacement against null hypotheses. *Evolution*, **33**, 897–913.

Strong, D. R., Simberloff, D., Abele, L. G., and Thistle, A. B. (eds.) (1984). *Ecological Communities: Conceptual Issues and the Evidence*. Princeton University Press, Princeton, New Jersey.

Taylor, R. J. (1984) *Predation*. Chapman & Hall, New York.

Tilling, S. M. (1985a) The effects of density and interspecific interaction on mortality in experimental populations of adult *Cepaea* (Held.). *Biol. J. Linn. Soc*., **24**, 61–70.

Tilling, S. M. (1985b) The effect of interspecific interaction on spatial distribution patterns in experimental populations of *Cepaea nemoralis* (L) and *C. hortensis* (Mull.). *Biol. J. Linn. Soc*., **24**, 71–81.

Tilman, D. (1982) *Resource Competition and Community Structure*. Princeton University Press, Princeton, New Jersey.

Townsend, C. R., and Hildrew, A. G. (1979) Resource partitioning by two freshwater invertebrate predators with contrasting foraging strategies. *J. Anim. Ecol*., **48**, 909–920.

Van Delden, W. (1984) The alcohol dehydrogenase polymorphism in *Drosophila melanogaster*, facts and problems. In: *Population Biology and Evolution*, Wöhrmann, K., and Loeschcke, V. (eds.) Springer-Verlag, Berlin.

Vandermeer, J. H. (1969) The competitive structure of communities: an experimental approach with Protozoa. *Ecology*, **50**, 362–371.

Varley, G. C., Gradwell, G. R., and Hassell, M. P. (1973) *Insect Population Ecology: An Analytic Approach*. Blackwell, Oxford.

Vigue, C. L., and Johnson, F. M. (1973) Isozyme variability in species of the genus *Drosophila*. VI. Frequency-property-environment relationships of allelic alcohol dehydrogenases in *D. melanogaster*., *Biochem. Genet*., **9**, 213–227.

Volterra, V. (1926) Variations and fluctuations of the number of individuals in animal species living together. Translation in: *Animal Ecology* by R. N. Chapman, McGraw-Hill, New York, 1931.

Westoby, M. (1984) The self-thinning rule. *Adv. Ecol. Res*., **14**, 167–225.

Whitehouse, H. L. K. (1973) *Towards an Understanding of the Mechanism of Heredity*, 3rd edn. Edward Arnold, London.

Williams, C. B. (1964) *Patterns in the Balance of Nature; and Related Problems in Quantitative Ecology*. Academic Press, London.

Williamson, M. (1957) An elementary theory of interspecific competition. *Nature*, **180**, 422–425.

Williamson, M. (1972) *The Analysis of Biological Populations*. Edward Arnold, London.

Williamson, M. (1981) *Island Populations*. Oxford University Press, Oxford.

Williamson, P., Cameron, R. A. D., and Carter, M. A. (1976) Population density affecting adult shell size of the snail *Cepaea nemoralis* L. *Nature*, **263**, 496–497.

Williamson, P. G. (1981) Palaeontological documentation of speciation in Cenozoic molluscs from Turkana Basin. *Nature*, **293**, 437–443.

Wilson, D. S. (1975) The adequacy of body size as a niche difference. *Am. Nat*., **109**, 769–784.

Yokoyama, S., and Schaal, B. A. (1985) A note on multiple-niche polymorphisms in plant populations. *Am. Nat*., **125**, 158–163.

Author Index

Subject Index